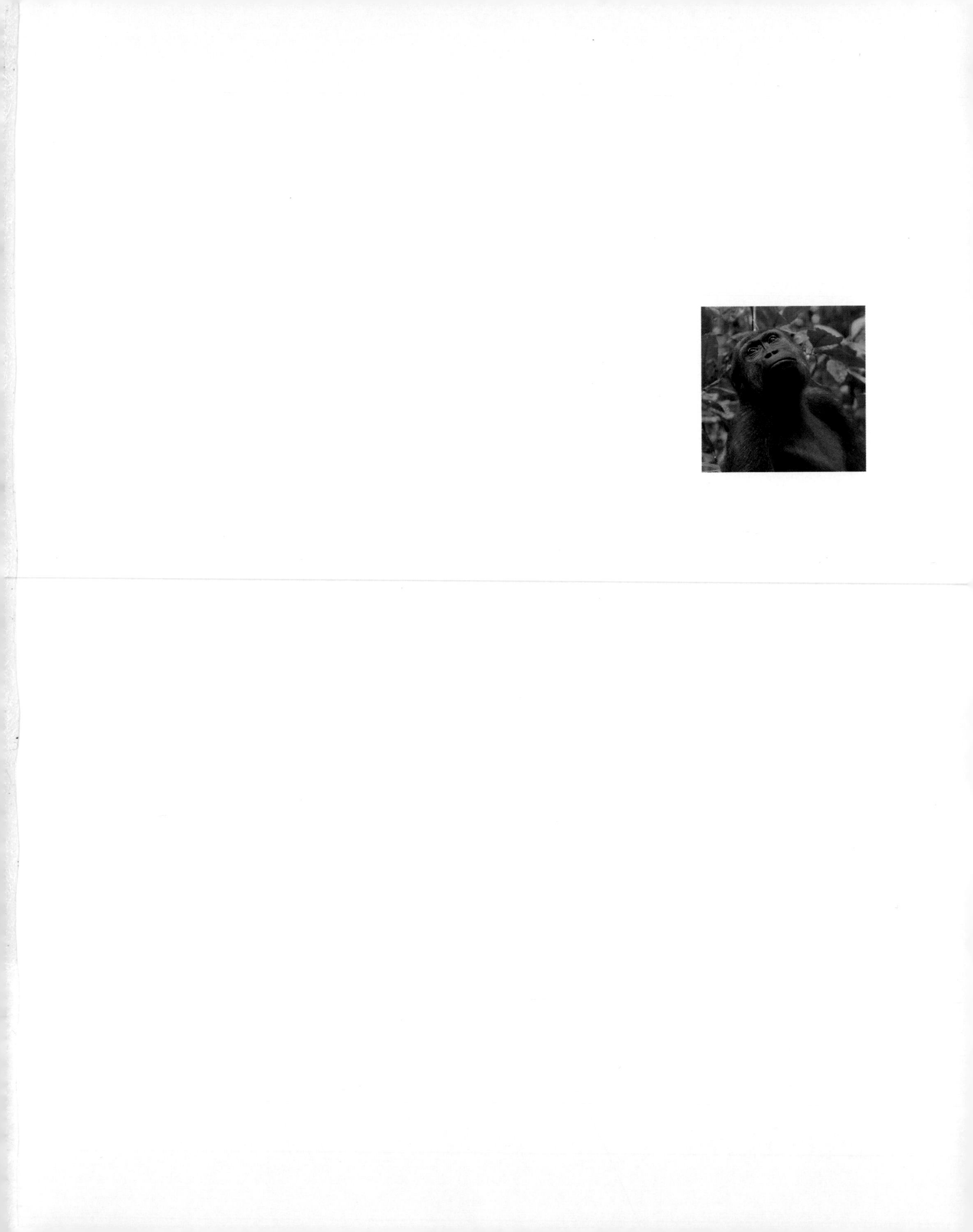

Gerardo Ceballos, Anne H. Ehrlich, and Paul R. Ehrlich

WITH ORIGINAL ART BY DING LI YONG

THE ANNIHILATION OF NATURE

Human Extinction of Birds and Mammals

JOHNS HOPKINS UNIVERSITY PRESS BALTIMORE

Printed in China on acid-free paper
9 8 7 6 5 4 3 2 1

Johns Hopkins University Press
2715 North Charles Street
Baltimore, Maryland 21218-4363
www.press.jhu.edu

Library of Congress Cataloging-in-Publication Data

Ceballos, Gerardo.
The annihilation of nature : human extinction of birds and mammals / Gerardo Ceballos, Anne H. Ehrlich, and Paul R. Ehrlich; with original art by Ding Li Yong.
pages cm
Includes bibliographical references and index.
ISBN 978-1-4214-1718-9 (hardcover : alk. paper) — ISBN 978-1-4214-1719-6 (electronic) — ISBN 1-4214-1718-9 (hardcover : alk. paper) — ISBN 1-4214-1719-7 (electronic) 1. Nature—Effect of human beings on. 2. Extinction (Biology). 3. Rare birds. 4. Rare mammals. 5. Biodiversity conservation. I. Ehrlich, Anne H. II. Ehrlich, Paul R. III. Title.
GF75.C43 2015
591.3'8—dc23 2014041105

A catalog record for this book is available from the British Library.

Special discounts are available for bulk purchases of this book. For more information, please contact Special Sales at 410-516-6936 or specialsales@press.jhu.edu.

Johns Hopkins University Press uses environmentally friendly book materials, including recycled text paper that is composed of at least 30 percent post-consumer waste, whenever possible.

In memory of our friend Navjot Sodhi, a brilliant conservation biologist
who started on this journey with us but could not finish.

It was the best of times, it was the worst of times,
it was the age of wisdom, it was the age of foolishness,
it was the epoch of belief, it was the epoch of incredulity,
it was the season of Light, it was the season of Darkness,
it was the spring of hope, it was the winter of despair,
we had everything before us, we had nothing before us.

Charles Dickens, *A Tale of Two Cities*

Contents

Preface

Only about fifty Tasmanian orange-bellied parrots still exist in the wild.

Humanity has unleashed a massive and escalating assault on all living things on this planet. The purpose of this book is to shine a spotlight on the onslaught, focusing on losses of the animals that are most familiar to people: birds and mammals. The roots of this destruction run deep through time. Human hunting and other activities were responsible for pushing populations of animals to extinction long before the agricultural revolution, which began about ten thousand years ago. Today, however, our collective assault on animals, plants, and microbes has reached such a horrendous level that any alarm call we might sound will be too faint to match the tragedy that is unfolding. The alarm must be amplified. Listen to our stories. Learn what's happening, and demand change.

In the past century or so, both the population size and technological capabilities of *Homo sapiens* have increased spectacularly, and this is the root cause of the rapid acceleration in human-caused species extinctions, precipitating what is now called the sixth great extinction wave. That label is based on powerful evidence that human actions may devastate the living world to a similar or even greater extent than did any of the five great extinction waves that demolished biodiversity, ending geological periods in distant prehistory. Those previous five extinction waves, which we describe in chapter 2, were unconnected to any supposed intelligent species; they were due to natural events that eliminated as much as 90 percent of Earth's flora and fauna in each occurrence. Each time, life recovered over many millions of years, producing a new array of life-forms. The last great extinction event occurred about 65 million years ago, long before the first upright, small-brained human ancestors moved onto the savannas of Africa.

The decimation of life-forms that is under way today is unlike the others because it is human induced and will have extremely serious consequences for humanity. The sixth extinction wave is the first to occur since the evolution of *Homo sapiens*. Because of humanity's success in populating and dominating the world, this new extinction spasm is proceeding rapidly and unfolding virtually everywhere. This new wave may also be a harbinger of the end of our global civilization, because the endangered populations and species of microbes, plants, and nonhuman animals are not just our only known living companions in the universe, but are working parts of life-support systems on which civilization depends.

The current extinction crisis and the risks it poses to humanity have been clear to scientists for a long time, but they are still almost always ignored, or even ridiculed, by others, including those who govern us. Vital statistics such as estimates of the number of species that have gone extinct or of the number of populations that have disappeared can be intellectually impressive. They don't, however, seem to carry the emotional impact to the public that the loss of biodiversity has for those of us who are deeply familiar with the impoverishment of nature. It is rather like the way, for most of us, that the many thousands of violent human deaths of strangers we hear about each year are regrettable, but they do not convey the emotional impact of the demise of a close friend or loved one.

To gain support for conservation initiatives and for maintaining the essential natural services on which humanity depends, it seems essential to communicate to the public and politicians the emotional side of the plight of biodiversity and to connect that predicament to human well-being. That, beyond giving the bare statistics, is what we attempt to do with this book on the endangerment and extinction of birds and mammals, the groups of animals with which most of us empathize.

We hope we can help you relate to the fate of the last wild Spix's macaw, a male that searched fruitlessly for a mate until it disappeared from the savanna of northeastern Brazil in 2000. We pray that we can make you care about that weird, wonderful, and poisonous mammal, the duck-billed platypus. We believe we can help you understand what a tragedy it would be if we lost the dwindling and little-known Cross River gorillas.

This book describes the horrors that have befallen too many of the world's birds and mammals and threatens too many more in a rapidly deteriorating situation. We want to familiarize you with the stories of our feathered and furry relatives—species that people have either expunged from Earth or are in the process of exterminating. We want to give you a feel for what interesting creatures they were and are, and recruit you to help stop an ongoing biotic genocide.

You needn't be a professional biologist to contribute. Indeed, in the 1930s the great American naturalist Aldo Leopold promoted the importance of wildlife research as a hobby for laypeople. More recently, a wonderful woman, Marj Andrews, in Queensland, Australia, has impressed biologists looking into the status of the Eungella honeyeater, a bird species described only thirty years ago, which has a tiny range. Marj, now in her eighties, had been making meticulous field notes on this and other species as the area around her was being logged, converted into farmland, and covered with strip mines. She produced an invaluable record of the biology of one interesting species and the fate of local birds under human pressure. Many amateurs like Marj are contributing to scientific knowledge about the fate of Earth's biodiversity, and we hope to recruit more into doing it.

But first, we'll try to describe the vast extent of Earth's macroscopic (visible) biodiversity—how many different kinds of plants and animals share the planet with us and make it hospitable to human life. We'll try to present this book with as little technical jargon as possible. We'll be consistent in our use of common names, while including a full list of the Latinized "scientific" specific names (each of two parts, a generic, or genus, name followed by the "trivial" name, as in *Homo sapiens*) in the appendix. Having the scientific names handy should make it easier for you to learn more about particular creatures. Internet searches for those names will work in most cases.

We also provide recommended reading with brief annotations for each chapter at the end of the book. We have given weights and measures in the metric system, but also include English measures in parentheses for the benefit of those of us educated in countries that still resist the tide of universal metric usage. In this book we use the word "discover" for species when they are first identified scientifically. Of course, many indigenous groups knew much about most of the species described here long before the first Western naturalist or explorer ever collected a specimen.

A special note: we have dedicated this book to our friend and colleague, Professor Navjot Sodhi, who was arguably the leading conservation biologist in southern Asia. Navjot started the book with us, but died suddenly at the age of 49, after making substantial contributions to this effort. We found finishing the book without his expertise and renowned sense of humor a sorrowful task. The world will miss his valiant efforts to save biodiversity. All royalties from publication will be donated to the Navjot Sodhi Memorial Fund, Rocky Mountain Biological Laboratory, Crested Butte, Colorado. Those funds will support the work of young conservation biologists.

Acknowledgments

We are extremely grateful to our friends and colleagues who helped us with discussion, comments, and reviews of parts of the book. We are particularly indebted to Anthony Barnoski, Daniel Blumstein, Rodolfo Dirzo, Michael Donohue, Daniel Karp, Rodrigo Medellin, Stuart Pimm, and Jai Ranganathan for commenting on early drafts of chapters and to three anonymous reviewers who helped us polish the manuscript. We are also grateful to Gretchen Daily, Jared Diamond, Rurik List, Jack Liu, Chase Mendenhall, Robert Pringle, Graham Pyke, and Jonathan Rossouw for comments on specific issues. Ding Li Yong kindly produced the illustrations of extinct species. Scott Altenbach, Claudio Contreras Koob, Peter Harrison, John Hessel, Jack Jeffrey, Frans Lanting, Susan McConnell, Alexander Pari, Roberto Quispe, Roland Seitre, Lynn M. Stone, and Jorge Urbán graciously allowed us to include their superb photos. Lourdes Martinez and Jesus Pacheco kindly helped us with the appendix and the index. Maria denBoer did a prompt and meticulous job of copyediting. The Universidad Nacional Autonoma de Mexico (DGAPA) has supported the work of Gerardo Ceballos, and for that he is most grateful. Anne and Paul Ehrlich are indebted to Peter and Helen Bing, Larry Condon, and Wren Wirth, who have made their work possible. Finally, we want to express our profound gratitude to our editor, Vincent Burke, for his care and help at every stage of this project.

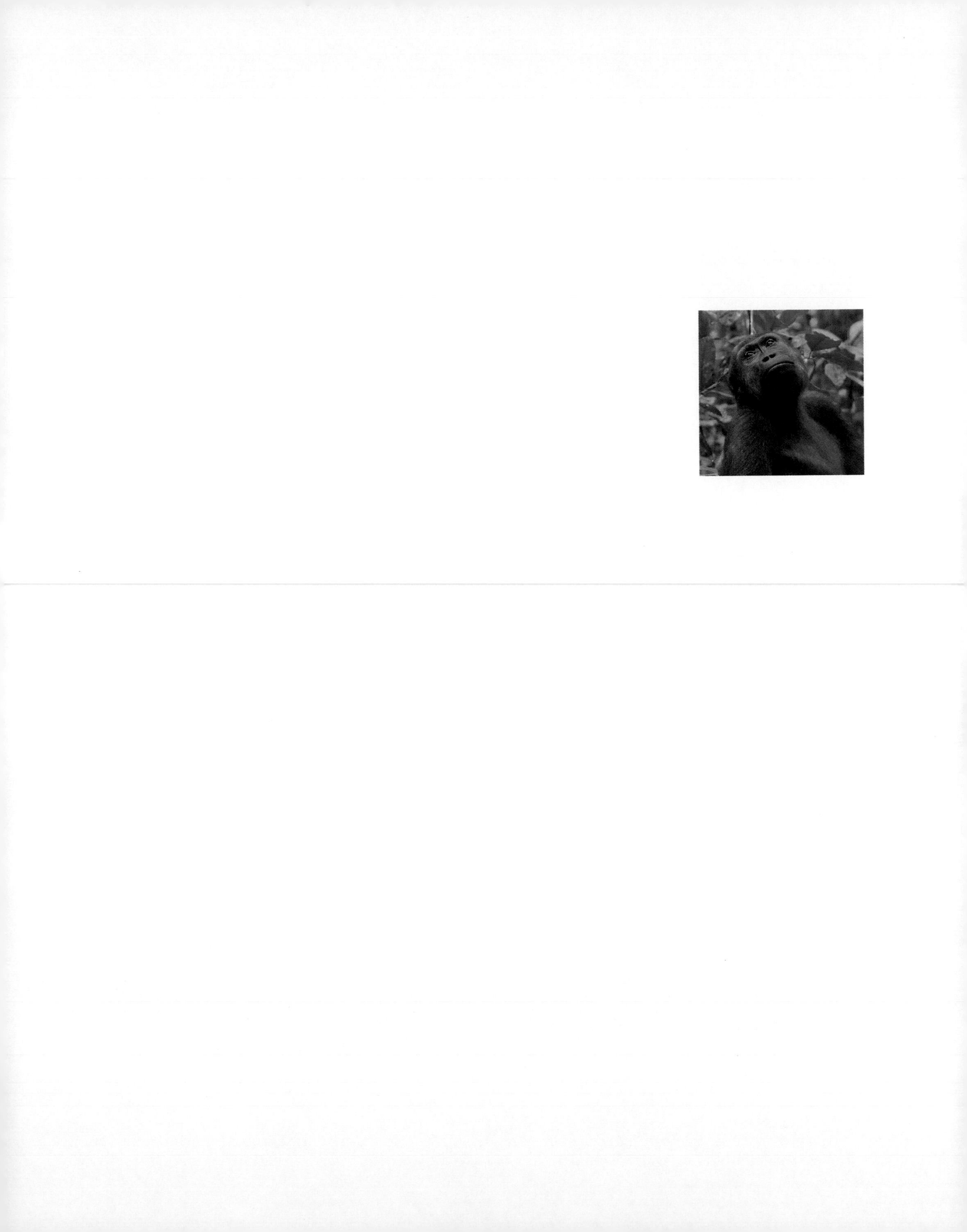

Tropical rain forests are the most diverse ecosystems on Earth. Millions of species of plants, animals, and microorganisms, most of them unknown to science, live in these complex and marvelous ecosystems, which are highly threatened by human activities.

1 THE LEGACY

ELSEWHERE IN OUR huge, cold, and little-comprehended universe, containing more stars than all the grains of sand on our planet, there may be life. But life might, as far as we know, be exclusive to Earth. Ours is a fascinating planet. It had its beginning roughly 4.6 billion years ago, when it condensed, in very complex cosmological processes, from interstellar dust and gas. One billion years later, microscopic life was well established in the oceans, having originated from still little-understood events.

Although people have been interested in ancient life since the time of the Greek philosophers, it was not until the mid-twentieth century that the technology to date very old fossils accurately was developed. Previously, the most precise date for the earliest life on Earth was estimated to be about 1 billion years ago. But a surprising discovery was made in 1983 in Warrawoona, in northwestern Australia. An insightful group of scientists found fossils of various kinds of filamentous bacteria approximately 3.5 billion years old in stromatolite beds. Stromatolites are ancient structures formed by the binding of sand grains by microorganisms, primarily cyanobacteria (blue-green algae). Life indeed began long, long ago. The earliest life, however, was possibly significantly older even than 3.5 billion years, as these scientists noted, because these bacteria were already pluricellular associations on their way to multicellularity, indicating that they had by then evolved significantly beyond their single-cell bacterial ancestors. Stromatolites can still be found in shallow coastal seawater in places such as Baja California and Western Australia.

The diversity of life has come a very long way since those tiny, little-known ancestral organisms evolved into millions of different species of plants, animals, fungi, and microorganisms. Today, tropical forests and coral reefs contain Earth's richest reservoirs of species, and the biological wealth in both is extraordinary. For example, a single hectare (2.47 acres) of a forest near Iquitos in Peru holds about 150 tree species, and 5 hectares (12 acres) in the forests of Borneo contain about 1,000 tree species. In contrast, in all of North America north of Mexico, a region covering almost 2 billion hectares (5 billion acres), there are fewer than 1,000 tree species. Thus, it is not surprising that the ideas of Charles Darwin and Alfred Russel Wallace (who both developed the theory that evolution is driven by natural selection) were inspired by the diversity of life they saw in tropical regions.

Animals are far less obvious in tropical forests than are plants, and seeing any large animal there is a matter of luck. Yet, in terms of diversity, invertebrate animal life in tropical forests is even more astounding than that of the vegetation. Invertebrates are animals without backbones and include shellfish, squid, octopuses, and myriad other creatures in the oceans. On land, the best-known group of invertebrates is insects. The diversity and abundance of tropical insects is legendary: a single tree in the Amazon region may host hundreds of species of beetles and more species of ants than in the whole of Great Britain. We are reminded of a famous quote attributed to the brilliant British scientist J. B. S. Haldane. When a theologian asked him what his study of biology had revealed

about the "mind of the creator," he replied that the creator must have had "an inordinate fondness for beetles." Insects, as Haldane's quote suggests, are one of the most diverse groups of animals, with more than a million known species and thousands more described by scientists every year.

As surprising as it may seem, we are far from having even a rough grasp of the total number of species of plants, animals, and microorganisms that populate our planet, even though roughly 1.8 million species have been named and described so far. Recent estimates of the total number of species on Earth have ranged from a few million to more than 100 million species. And in the past few years, some researchers have even estimated the number of species in the billions. Much, of course, depends on how one defines "species" and how microorganisms (including viruses) are evaluated, which is still the topic of long arguments among scientists.

Surprisingly, more than five hundred species of mammals have been discovered and described by scientists in the past decade. We are in a "New Age of Discovery," where previously unknown organisms are being brought to light across the world. Unfortunately, many of these species are found in regions wracked by habitat destruction and are themselves highly threatened. Two examples of such discoveries are a new night monkey (*top*) and an "extinct" Inca rat (*bottom*).

Regardless of the definition of a species and the exact number of them, the diversity of living organisms is truly astounding. Contrary to what most people believe, discoveries of new species are still quite common, especially among plants, invertebrates, and microorganisms. A recent survey by the International Institute for Species Exploration at Arizona State University reported that 18,516 previously unknown species were discovered in 2007, amounting to an average of 50 discoveries a day and equivalent to roughly 1 percent of all described species. Similarly, the results of the first Census for Marine Life were announced at the end of 2010, a decade after it was launched. The scientists involved described more than 1,200 marine species, with 5,000 additional possible species still in the process of being described. Another decade-long effort in the Mekong River basin uncovered more than 1,000 new species of plants and animals. From 1999 to 2009, scientists working on the Amazon Alive project in the Amazon basin discovered a wealth of new species: 637 plants, 257 fishes, 216 amphibians, 55 reptiles, 16 birds, 39 mammals, and thousands of invertebrates such as insects, spiders, and worms.

Thousands of life-forms have also been discovered in the most unusual places. Thermophilic (heat-loving) bacteria have been found thriving at temperatures up to 140 degrees Celsius (284 degrees Fahrenheit), well above the boiling point of water. They live in places such as geysers in Yellowstone National Park (United States) and hydrothermal vents in the floor of the Pacific Ocean. The discovery of these bacteria challenged the view that organisms could not survive at high temperatures because their proteins would denature—lose their shape and function—something that often occurs at high temperature. Discovering the physiological mechanisms by which these bacteria maintain their proteins intact at high temperatures may lead to finding a way to prevent people from dying from high fevers. And other remarkable bacteria, microorganisms, and invertebrates have been discovered in places such as the barren rocks of Antarctica, the ocean floor thousands of meters below the surface, and 1.3 kilometers (0.8 mile) deep in mines, to mention a few odd places where life has been found. In addition, the oceans themselves appear to be replete with viruses.

The discovery of new species even among mammals and birds is surprisingly frequent, occurring throughout the world. Most discoveries, however, have been made in tropical regions, where there are many key sites for newly found species. The tropical forests of Southeast Asia and Oceania, extending from Burma to Vietnam and Cambodia, and islands such as Borneo and Papua New Guinea, have harbored many remarkable discoveries. In Africa, key sites include the mountains and forests of Kenya and Tanzania, the Congo basin, and the island of Madagascar. And in the Americas, scientists working in the Amazon basin and the Andes foothills have uncovered many interesting species.

Our own scientific work has shown that almost 10 percent of all mammal species—more than four hundred—have been scientifically described just since 1993. Newly discovered mammals, as one would expect, are mostly small creatures such as mice, shrews, and bats. Yet the list also encompasses more than sixty species of monkeys, including a golden capuchin found on a sugar plantation in an extensive agricultural landscape in the northeastern Brazilian state of Pernambuco. The entire population of the capuchin consisted of 18 individuals and was restricted to a 200-hectare (500-acre) forest remnant surrounded by sugar cane fields. Other new species of monkeys, including a silver marmoset, have been discovered in Brazil during the past decade.

In India the Arunachal macaque, another primate, was described in 2004 living in the Himalayan foothills at elevations up to 3,500 meters (11,500

Forest-dwelling, nocturnal mouse lemurs are, like the aye-aye and all other lemurs, confined to Madagascar. Even with tails about 25 centimeters (10 inches) long, they are the smallest living members of our own order, Primates. Their classification is in flux, and many new species have been discovered since 2000. Madagascar is one of the most ecologically devastated areas on Earth, so it is no surprise that habitat destruction is moving many lemurs toward extinction.

feet), one of the highest elevations at which primates are known to live. An entirely new monkey genus and species, the kipunji, was discovered in the Rungwe Mountains in Tanzania in 2007; it was the first new monkey genus to be found in eighty-three years. Primate species discovered since 2010 include the Caquetá titi monkey from rain forests in Colombia; the northern buffed-cheeked gibbon from the rain forests of the Annamite Mountains around the borders of Vietnam, Laos, and Cambodia; and three species of slow lorises, beautiful nocturnal animals that are the only primates with a poisonous bite. They secrete a toxin in glands located in the elbows, which they lick and mix with saliva.

New species of whales, flying foxes, shrews, gerbils, pygmy sloths, otters, civets, and many other kinds of mammals have recently been found in remote and not-so-remote places around the globe. One of the most remarkable discoveries was of members of a family of rodents similar to squirrels. The first live specimens known to Western science were bought at a local market in Laos, where they were being sold as food. The new species was a member of the rodent family Diatomyidae, previously known only from 11-million-year-old fossils. This is an example of what is called the Lazarus effect, in which an organism known only from fossils is discovered to be alive.

A most remarkable discovery in the twenty-first

century has been a new tapir, recently found in the Kabomani region in the Amazon basin between Brazil and Colombia. It is the first tapir described since 1865. It is also the first species of the order Perissodactyla to be described in one hundred years. The new Kabomani tapir is distinct from other tapirs, being black and much smaller, weighing around 110 kilograms (240 pounds), compared to the 320-kilogram (710-pound) Brazilian tapir.

Recent genetic studies have revealed further surprises, raising the number of known mammals even higher. For example, forest and savanna elephants in Africa are now considered different species, as are the orangutans of Sumatra and Borneo and clouded leopards on the Southeast Asian mainland and in Borneo and Sumatra. Indeed, the diversity of mammals may actually be around 8,000 species, instead of the 5,500 presently acknowledged.

Birds, being a bit more obvious and the focus of attention of both ornithologists and bird watchers, are certainly better monitored than mammals. Thus we have a more dependable estimate of the number of bird species alive today, around 10,000. Nevertheless, roughly 100 new species have been described since 1991. In a fascinating exploration of the Amazon forest in Brazil, scientists located two new bird species, the bald parrot (also called the orange-headed parrot) and the cryptic forest falcon. The isolated Santa Marta mountain range of Colombia has an astounding thirty-two species of endemic birds, many just discovered or rediscovered. The Santa Marta Mountains are

Scientists recently discovered that the forest elephant is more taxonomically and ecologically distinct from the familiar bush elephant than previously thought. Even though it may weigh 3 tons (2.7 metric tons) or more, it is the smallest of three extant elephant species. Forest elephants are in dramatic decline in their Central African homeland, some 60 percent of the population having been lost in the first decade of the twenty-first century. They suffer from poaching for ivory, killing for bushmeat, poor enforcement of laws to protect them, and habitat loss in the face of burgeoning human populations. In May 2013 twenty-six forest elephants were slaughtered in the Dzanga clearing, famous as a daily gathering place for the elephants. Their ivory went to finance a rebel group's military operations.

The Santa Marta mountain tanager is found only in the Santa Marta Mountains of Colombia. At the moment it is not considered endangered, but its small geographic range should keep people alert to its status. Tanagers are a group of several hundred species in the Western Hemisphere, and the great beauty of many of the males makes them an important draw for ecotourism. They are still common in much of tropical America, often occurring in mixed-species flocks feeding on fruits, seeds, and insects.

completely cut off from other mountains by a sea of lowland tropical forest.

In India, an astronomer who was an amateur birder found the first new bird discovered in more than fifty years in that country. His discovery was that of a small, colorful babbler, known as the Bugun liocichla, in the forests of the Eaglenest Wildlife Sanctuary in the state of Arunachal Pradesh, close to India's border with China. He first observed the bird in 1995, but it was not until 2006 that the species was officially recognized. This newly described species may not be around long, as the only three known populations consist of an estimated two hundred individuals.

The colorful babbler and all the other species on this planet are a legacy. They have been handed down to us as surely as the valuables described in a will are handed down by an ancestor. And like any inheritance, whether the gifts will be cherished or wasted depends entirely on what the recipient does with them. If our past is our future, waste will win. If we continue the course of destruction then, as the famed biologist E. O. Wilson pointed out, we are engaged in a folly for which our descendants will never forgive us.

But our destruction of the legacy of biodiversity is not predetermined. In this book we showcase the two most popular ambassadors of biodiversity, birds and mammals, in a way that clarifies the paths we have taken and the options before us, whether toward conservation or heedless annihilation. It is our hope that by the end of this book you will feel so connected to these birds and mammals—and so outraged by what has been done to them—that fate will change its course and the unfolding mass extinction event will be aborted. The images we have chosen and the language we use are intended to provoke your thoughts and feelings, to stir within you a need to act. We ourselves have seen many of these magnificent creatures in their native habitats, sometimes returning years later to find the birds gone, the mammals vanquished, and the paradises ruined. Through this book we are sharing with you the best and worst experiences, hoping that we can convey our own feelings of sadness, loss, frustration, fascination, and hope, so that our pill, both bitter and sweet, might be a cure.

A thunderstorm in the desert of Baja California in northern Mexico. It is a reminder of the strong forces of nature that have shaped life on Earth over billions of years. Major natural cataclysms were the cause of the first five catastrophic mass extinctions. Now we are the cause of the sixth.

2 NATURAL EXTINCTIONS

An extraordinary planetary event occurred on March 22, 1989, when an asteroid three times as wide as a football field just missed hitting Earth. The asteroid passed through the exact spot where Earth had been only a few hours before. But for that near miss, the planet would have felt an impact similar to exploding 1,000 to 2,500 1-megaton hydrogen bombs simultaneously. That movement made the difference between a brief evening news story and the immediate deaths of millions of people, not to mention enormous damage to both human infrastructure and biological diversity. The result would have been the most recent natural mass extinction episode. Much of the planet's surface would have been, for a long while afterward, a wasteland.

Life on Earth in other times has been less lucky. Paleontologists divide Earth's history into periods marked by transitions that produced abrupt changes in the species composition of fossils. Thus the division between the Precambrian and Cambrian periods, 600 million years ago, is characterized by an abrupt shift from only microscopic marine organisms to the addition of many complex visible life-forms such as trilobites, now-extinct marine relatives of today's insects and crabs. These transitions have been associated with natural disasters that caused catastrophic, widespread perturbations wherein major groups of plants and animals became extinct over a relatively short time, changing the course of evolution.

These devastating events are called mass extinctions. The five largest mass extinction waves occurred during the past 500 million years; all unfolded rapidly (in geological terms) from natural causes, such as extraordinarily extensive and prolonged volcanic activity inducing global climate cooling, or rapidly warming periods, or asteroid impacts. Yet the effects were not uniform; some large groups of related species were lost, while some others remained largely unaffected. Each time, however, the resulting catastrophic loss of global biodiversity required the Earth to wait millions of years to re-establish a similar abundance.

The most recent mass extinction event was the Cretaceous–Tertiary transition (often called the K-T boundary), which occurred around 65 million years ago and wiped out almost all of the dinosaurs, which had previously dominated the planet. Within a few thousand years, an estimated 95 percent of all previously existing species disappeared, an extinction event of gigantic proportions. Of the dinosaurs, only the ancestors of the birds survived, a legacy that enchants us today.

Debates about the K-T boundary mass extinction may never end, but it is generally attributed to the collision of our planet with a massive asteroid that hit the Yucatán Peninsula at the edge of the Gulf of Mexico. Watching the calm waves of the Gulf today, one would have a hard time imagining the brutal force of an ancient collision that vaporized all life at ground zero. The asteroid likely passed through the atmosphere in about 1 second, heating the air in front of it to several times the temperature of the sun. When it hit, the asteroid itself would have vaporized, launching an enormous fireball into space. Rock particles would have been flung thousands of miles into space. Gigan-

tic shock waves would have reverberated through the bedrock and then returned toward the surface to send melted globs of rock partway to the moon.

Some of what went up would have come back down, and returning debris would have rained down trillions of speeding meteors, setting myriad forest fires. These conflagrations would have covered huge areas of the continents and burned for weeks or months. Earthquakes and resultant landslides and tsunamis would have added to the devastation. Whether or not this description is accurate, all land animal species weighing more than 18 kilograms (40 pounds) became extinct, including all the dinosaurs except the birds. But a group of inconspicuous small animals at the time, the mammals, with their weird-looking, hair-covered bodies, pulled through. In time, over millions of years, the mammals became one of the most successful groups on Earth, diversifying into the thousands of species we know today.

Of course, extinctions were occurring all the time between the mass extinction episodes. New species

The brown bear (called the grizzly in North America) is not in danger of global extinction at the present time. It once ranged from Mexico to Alaska and from northern Europe and Russia to North Africa, and is still found in parts of North America (where it can be abundant) and northern Eurasia. But the bear's range has contracted greatly; its Russian population has been cut in half in two decades, and many smaller populations are under assault from the usual culprits, in some places including hunting for paws and gall bladders. These cubs, like our own grandchildren, face a very uncertain future.

Grevy's zebras were once relatively widespread, but now have been mostly confined to the Horn of Africa (including parts of Kenya and Ethiopia). An extreme decline in this beautiful relative of the common zebra is attributable largely to reduced access to food and water because of competition with exploding populations of pastoral people and their animals. Hunting for hides, meat, and the imagined medical value of some body parts has also contributed.

are continuously formed as natural selection works on populations in changing environments, and populations are continually going extinct in response to environmental changes. When the last population of a species goes extinct, the species goes with it. All that is perfectly natural, as Earth has always been in a process of gradual change. Plate tectonics move the continents around, causing their climates to change, and volcanic activity can alter the climate of the entire planet. Mountains erode, glaciers come and go, new land is created, and old land sinks beneath the seas. New species evolve, novel kinds of predators coevolve with plant-eaters, and opponents that fail to adapt disappear. Flux is the natural state of the world, but usually the changes are so slow that most human individuals don't even notice them.

From space, Earth now seems to be a quiet planet, only sporadically shaken by a volcanic eruption or an earthquake and tsunami. Nevertheless, the rate of change is accelerating, increasing the rate of population and species extinctions and vastly outpacing the compensating rates of population differentiation and speciation. Hundreds of species of mammals, birds, and other vertebrates have been recorded as having become extinct in the past five hundred years. A plethora of additional overlooked species, including invertebrates and plants, has likely become extinct without a record. And millions of populations and species today are facing probable extinction.

Our planet is now entering a cataclysm so enormous that the diversity of life itself, with all its astonishing and wonderful animals, plants, and microorganisms, and the interactions among them, is again imperiled. But this time it is not threatened by cosmic or geologic forces, but entirely as the result of the activities of our species. While the fate of countless species hinges on what human beings do in the next two or three decades, paradoxically, civilization's survival depends on their fate.

A mountain gorilla in Rwanda. The picture answers the classic question, "Where does a full-grown male 'silverback' gorilla sleep?" Answer: "Anywhere he wants to!" But despite the saying, unless annoyed, gorillas are peaceful vegetarians; and even when bothered, they (like chimpanzees) almost never hurt people. A world without wild populations of these close relatives of ours would be a sad place indeed.

3 THE ANTHROPOCENE

The rise of human beings has critical implications for the future of life on Earth. For billions of years, biodiversity had been assaulted on a large scale by cataclysms that occurred at intervals of tens to hundreds of millions of years. Then modern *Homo sapiens* came along. Over many thousands of years, people went from being rock-throwers to users of traps, spears, and bows and arrows and, eventually, gun-wielding hunters. Along the way, behaving much like other omnivorous predators, we fed and clothed ourselves by killing and eating plants and other animals. But our efficiency was astounding, and long before inventing guns, we probably exterminated great herds of medium-sized and large herbivorous mammals such as mastodons, ground sloths, and several species of giant kangaroos. Without a shot being fired, we also wiped out threats to our safety such as cave bears, dire wolves, and a marsupial carnivore as big as a lioness.

In Africa, Europe, and Asia, where animals had long evolutionary experience with humans, large game animals mostly survived. But the relatively recent human invasions of Australia and the Americas seem to have contributed to numerous extinctions. Later, on some large islands, early human invaders found and quickly eliminated huge avians such as the elephant bird of Madagascar, a strange-looking, massive animal roughly 3 meters (10 feet) tall and about 400 kilograms (880 pounds) in weight, and the equally impressive moas of New Zealand.

Since the agricultural revolution began some ten thousand years ago, the human population has grown enormously. Incredible as it may seem, the number of people has increased more in the past one hundred years than in all the previous history of humanity. In 1930, the entire human population was 2 billion. By 1960, the population had increased to 3 billion. Today there are more than 7 billion people, and the number is growing by some 235,000 a day. Demographers project that the global population may reach around 9 billion in 2045 and more than 11 billion by 2100.

With regard to the planet's biodiversity, the most obvious correlation is this: more people equals fewer species. All organisms, including people, must be able to extract critical resources from their surroundings, release wastes into their environments, and have room to carry out these activities. As human populations explode, with some exceptions, the living space for other organisms, resources, and waste sinks (natural reservoirs that absorb or break down and recycle byproducts or unusable remains) decline. Major exceptions are those organisms that human beings have domesticated for their use, such as chickens, pigs, cattle, horses, and some varieties of grain and vegetables, or those that have learned to make their livings alongside humanity. Rats, human lice, dengue virus, and various weeds are good examples of the latter category. The consequences of the recent human population expansion dwarf the Pleistocene overkill, during which it is hypothesized that the spread of *Homo sapiens* around the world from 50,000 to 12,000 years ago killed off many large animals, most recently in Australia and the Americas. Today's sixth mass extinction wave extends from the twentieth into the twenty-first century, a period some scientists have called the Anthropocene.

Elephant birds were apparently numerous in Madagascar before humans arrived. Many egg remains, including whole eggs, have been found in some locations. The huge eggs have a volume more than 150 times larger than that of a chicken egg! Here an Antandroy tribesman holds a fossilized elephant bird egg in southern Madagascar.

This term recognizes that a huge and growing human population has become the principal force shaping the biosphere (the surface shell of the planet's land, oceans, and atmosphere, and the life they support).

During the Anthropocene, changing proportions of the atmosphere's gases are bringing about an end to the favorable climate people have enjoyed for ten millennia. That is the climate in which agriculture and civilization were able to develop. The Anthropocene has also seen dramatic alterations of the land surface and the oceans, especially in the past century. In some ways, the changes are unlike those the planet had previously experienced, and in other ways they are reminiscent of past cataclysmic events. Just as that asteroid 65 million years ago is thought to have devastated life, we may end up wiping out 95 percent of the world's species—along with ourselves.

The claim that people are causing Earth's sixth great extinction event is easy to support. The consequences for other life-forms of the alteration and fragmentation of habitats, overkill by hunting or harvesting, rampant pollution, the introduction of invasive species and diseases, gluttonous overconsumption, and now climate disruption are obvious to anyone who is willing to examine the data with an open mind. How bad is it? Scientists who work on conservation

The resplendent quetzal, a spectacular bird from Mexico and Central America, was once known as the flying serpent. The Aztec culture in ancient Mexico and other Mesoamerican groups considered it a deity, associated with the "snake god" Quetzalcoatl, because when the males fly, their very long, green tails resemble a snake. It is the national bird of Guatemala, appearing on its coat of arms; quetzal is the name of Guatemala's currency. Habitat loss is causing the disappearance of many populations of this species.

issues have concluded that every year millions of populations and thousands of species are being wiped out worldwide.

Why should we care about these disappearances if there are billions of uncatalogued populations and thousands of species of plants and animals known to science, and many more that have yet to be described? What does a cup of water matter when you have a full bucket?

First, human-caused extinction is catastrophic for ethical reasons. Every species is a unique entity, a product of billions of years of evolution. Once extinct, it is gone forever; the universe is unlikely ever again to see that particular collection of genes. Moreover, the loss of any species can lead to losses of others that coexist with it. The plants, animals, and microbes that live in a given area interact with each other and with their physical environments, creating and maintaining the conditions necessary to sustain life for them all. Removing any one species will have consequences for others.

Second, life generates our most essential natural resource: oxygen. Oxygen is a biological creation, which would be unavailable without the aid of nonhuman species. Plants and some microbes capture energy from the sun in the complex biochemical process of photosynthesis, producing energy-rich carbohydrate chemicals, and in the course of doing so release oxygen into the atmosphere. We and all other animals (as well as the plants that produce it) need that oxygen to obtain energy by slowly burning (oxidizing) the carbohydrates that plants manufacture and we eat. When we mess with the elements of ecosystems that provide that oxygen, we may be literally threatening the lives of our distant descendants. There's a lot of oxygen in the atmosphere now, but over the very long term it could be depleted.

Third, plant and animal populations are all tightly tied together in ecological feeding sequences, such as grass→cow→person→mosquito→bat, called food chains. Of course, numerous additional species (such as mice and insects that also eat the grass) are involved and connected to each other. Energy and nutrients are passed from one species of living organism to another through these linkages, which in turn are interwoven into complex food webs. Then, when death intervenes or plants lose leaves or animals shed skin and eliminate liquid and solid wastes, other organisms called decomposers make a living gaining energy from the carbohydrates, fats, and other complex organic

Eurasian bee-eaters are widespread predators on large insects, especially bees and wasps. Bee-eater species are generally doing reasonably well at the start of the sixth great extinction, but they already suffer from hunting for food and sport, from canalization of rivers that destroys the banks in which they make nest burrows, and potentially from reductions in the population sizes of their insect prey due to broadcast use of pesticides.

chemicals in corpses and their leavings. They convert these substances back to simpler inorganic chemicals such as nitrogen, potassium, phosphorous, and many trace minerals that can once again be recycled by various species into more life. When we remove many species, we disrupt this cyclical process of life, death, and renewal.

These three reasons seem to us compelling and can be summarized as "protect Earth's ecosystems or die." We all live in and with ecosystems, defined units containing life-forms, which come in many sizes. Your mouth is an ecosystem containing billions of bacteria, fungi, viruses, and other microbes, some helpful to your health, some harmful. Your ecosystem runs on energy that originates in the sun and that you get from the plants and (indirectly) the animals that you eat. The Amazon basin is a much larger ecosystem, but it operates under similar principles. Like almost all ecosystems, including your mouth, the Amazon is powered by the sun and contains countless organisms, including people. Energy travels through ecosystems; without it, they would quickly disintegrate. With a steady input of energy, life-forms can flourish and decomposers can recycle the materials that all life needs.

Human beings, like all organisms, are completely dependent on the functioning of the world's ecosystems, even if many of us would prefer to think otherwise or don't want to think about it at all. At no cost, natural ecosystems perform for us a vast range of essential functions, including maintaining a breathable mixture of gases in the atmosphere, supplying fresh water and controlling floods, generating and replenishing soils, disposing of wastes, pollinating crops and protecting them from pests, supplying wild fish and game, growing edible and medicinal plants, and keeping us from becoming neurotic basket cases by providing places for recreation and reflection. Finally, human-managed ecosystems, especially farms, provide their critical benefits with the help of the natural ecosystems in which they are embedded.

Sometimes the loss of even a single species can have profound impacts on an ecosystem. These so-called keystone species are life-forms that have a larger impact on the ecosystem than might be predicted by their abundance. For example, black-tailed prairie dogs are large ground squirrels that live in colonies of thousands (hundreds of thousands in the past) in grasslands in North America. They feed on grasses and herbs and create complex underground burrow systems. Their activities are very beneficial because they destroy seeds, seedlings, and small plants of invasive desert shrubs such as mesquite that, uncontrolled, would eventually turn arid grasslands into desert scrubland. The burrow systems also provide shelter and protection for a plethora of animals from insects to skunks, and help aerate soils and increase water infiltration, promoting soil fertility and preventing runoff and flooding.

Studies on prairie dogs in the southwestern United States, begun in the early 1990s when they were considered pests to grazing cattle, demonstrated that prairie dogs instead are essential to maintaining their grassland ecosystem's productivity. The disappearance of prairie dogs led to an invasion of scrubland vegeta-

tion and desertification, largely destroying the land's value for grazing.

The ecosystem services that humanity depends on are delivered locally by populations of species. Thus, it is just as important to worry about the high rate of destruction of populations as it is to worry about the loss of species. After all, if a species of insect-eating bat is decimated and only one population survives (some bat species are now under such a threat), no species extinction will have occurred, but people in areas that have lost bat populations might suffer more from mosquito bites and the diseases these pestiferous insects spread.

Usually, many populations of a widespread species will go extinct before the species finally disappears. So many passenger pigeon populations had been wiped out by around 1880 that it was no longer commercially profitable to hunt them, but the species hung on for another thirty years or so, ultimately dying out because it could no longer assemble the gigantic flocks required for successful breeding. One of the possible consequences of the loss of billions of those birds was an

Black-tailed prairie dogs were once one of the most abundant mammals on Earth, with a population estimated in the billions. They are fundamental in maintaining their grassland ecosystem and providing ecosystem services such as soil carbon storage and groundwater recharge. Nonetheless, they are still poisoned throughout their range and are susceptible to introduced diseases such as bubonic plague.

upsurge in the food supply of the deer mice that once competed with them. The resultant mouse population explosion was bad for people because the mice are a principal reservoir for Lyme disease, a serious bacterial disease that is transmitted to humans via a tick bite.

In overpopulated Rwanda, the naked eroded hillsides are reflected in the resulting muddy-red color of the rivers that are washing away the nation's agricultural soils. Rwanda's biodiversity is largely confined to a few small reserves, under constant assault by poor people desperate to extract wood and bushmeat or to establish small farm plots. Since the horrifying genocide in 1994, trees have been planted on every scrap of land not devoted to crops, roads, or buildings. They are mostly exotic eucalyptus, capable of providing some ecosystem services, but nearly useless for rebuilding the biodiversity of ravaged Rwandan landscapes. Yet those landscapes might look idyllic to tourists lacking an ecologist's perspective.

Different human activities often create synergistic effects (ones whose combined impact is much greater than the sum of their individual impacts). A common example is when human activities greatly restrict the distribution or range of a species and also alter the climate locally or regionally. Earth is dotted with species newly rare because of human activities and thus more vulnerable than their widespread ancestors to further climate disruption, natural disasters, or other human actions. These "zombies" are the living dead. They might have survived either range restriction or climate change, but they will not survive both.

While it is true that extinctions of populations and species are natural processes and have been present throughout the history of life, it's also important to remember that today we are seeing an extraordinarily high rate of extinctions, especially compared with the rate of evolution of new species. Sadly, there are not enough professional biologists or even amateur observers to begin to document all of the extinctions of species now occurring, even of well-known groups of organisms such as birds and mammals. And the task of recording the extirpation of populations is even more gigantic. Accordingly, some people claim that the seriousness of the extinction crises is overesti-

The vast Serengeti Plains of East Africa contain the largest biomass (living weight) of mammals remaining on Earth. The name in the Maasai language means, appropriately, "endless plains." Visiting the Serengeti allows one essentially to look back into the Pleistocene, when such aggregations of large animals were much more common. Serengeti tourism is an important factor in the economies of Kenya and Tanzania, but the vast migrations are now threatened by plans for road-building in service of mineral development.

mated—intellectually the equivalent of claiming that a beach eroding away before one's eyes isn't disappearing because the number of grains of sand on it, or the number of grains disappearing, haven't been counted. Creating such doubts is as perilous as it is dishonest; if the current extinction wave continues to accelerate, it could be a harbinger not just of a tragic decline of the variety of life, but of the end of human civilization and the premature demise of billions of people. As the American cartoon character Pogo famously said, "We have met the enemy and he is us." Clever as we are, even understanding ecosystems as well as we do, we nonetheless behave in some very foolish ways. Whether from an ethical or a self-interested perspective, we surely can see that the road we are traveling is the wrong one. And yet we roar along, pedal to the metal, like Tolstoy's Muscovites with Napoleon at the gate, gaily dancing as destruction nears.

Guadalupe Island, off the Baja California coast in Mexico, was considered a biological paradise in the mid-1800s. Unfortunately, like on many other islands in all oceans, introduced goats, cats, and rats decimated its endemic birds and plants. Successful eradication efforts have allowed the recovery of some of its plants and animals.

4 LONG-SILENCED SONGS

A VISIT TO the California Academy of Sciences, located in San Francisco's Golden Gate Park, speaks volumes about the disaster that has befallen birds with the spread of humanity. A maze of narrow corridors in the scientific collections leads an explorer to the Ornithological Collection. There you will find a cabinet with a sign: "Extinct Birds." If you look inside, you'll experience a dreadful moment as you take in the sight of specimens of species that no longer exist. Your eyes will move from the imperial woodpecker and the passenger pigeon to the Guadalupe Island petrel, among many others. Each is carefully preserved in death—an ironic twist. We failed to preserve them while they lived, not so very long ago. Dread fades to sadness as your eyes linger on these inert specimens, the last samples of what once were animated creatures. Numbers have a way of numbing the mind, but seeing the remains of so many lost species of birds should touch a nerve and prompt a pledge.

Out of Africa

Until perhaps 60,000 years ago, only the birds and other animals of Africa would have encountered humans. The thousands of bird species in the Western Hemisphere, in most of Asia, in Australia, and on oceanic islands inhabited lands and waters untouched by modern human beings. Even in Africa, humans were few in number and lived on only some parts of the continent. But then they began to spread around the world, first moving into Eurasia, then occupying some southwestern Pacific islands and Australia, and finally invading the Western Hemisphere. Birds were confronted with a species unlike any that had preceded it. These smart, social, and hungry primates—our ancestors—subjected birds to hunting "blitzkriegs." In the face of that early version of "lightning warfare," many naïve birds never had a chance. The human diaspora was so fast in evolutionary terms that there was scant opportunity for birds valued for food or feathers to evolve anti-predator responses to our ancestors' onslaught.

Hunting was but one aspect of the human takeover. Unprecedented changes in habitats and accidental or deliberate introductions of animals such as cats, rats, goats, and pigs often caused more destruction than killing birds directly. As many as 2,000 bird species were driven to extinction on Pacific islands alone after human settlement began between 3,000 and 1,000 years ago, judging by sub-fossil remains and other evidence. Many of those birds were at a special disadvantage; lacking terrestrial predators, they had evolutionarily disposed of functional wings and so were unable to escape people or rats by flying. Furthermore, winged individuals were more likely to be blown off islands to their doom. The birds that did not maintain metabolically expensive and damage-prone functional wings in the millennia before *Homo sapiens* arrived out-reproduced those putting energy into then-useless appendages.

More recently, at least 132 bird species have gone extinct in these islands since the year 1500. Another fifteen are possibly extinct now, and four more species survive only in captivity. Many other avian extinctions have occurred across all continents, such as the pink-headed duck and Himalayan quail from India, the white-eyed river martin from Thailand, Bachman's warbler from the United States, the slender-billed grackle from Mexico, and the Atitlan grebe from Guatemala. Today, large numbers of endangered species are found in tropical regions such as Madagascar, Australia, the Philippines, India, China, Southeast Asia (including Vietnam and Cambodia), Sumatra and Borneo in Indonesia, eastern Brazil, and the Andes in Ecuador and Peru, where habitat loss and other causes have pushed countless species to the brink of extinction. The United States, Mexico, Brazil, Egypt, Tanzania, Angola, and South Africa also have large numbers of endangered species and populations. It is likely that the actual number of extinctions is even larger, as many species doubtless disappeared before they were known to science, while some others are extremely rare and have not been seen in such a long time that they are probably gone.

The Enigmatic Dodo

Among birds that went extinct early, the dodo has become the icon of extinctions. It was a ridiculously awkward-looking bird, found only on the island of Mauritius in the Indian Ocean, approximately 870 kilometers (540 miles) east of Madagascar. A Dutch sailor named Heyndrick Dircksz Jolinck, who visited the island in 1598, first described the dodo. He wrote of a large bird with pigeon-sized wings that were useless for flying. He noted, "these particular birds have a stomach so large it could provide two men with a tasty meal and was actually the most delicious part of the bird." But other contemporary reports of dodos indicated that their meat tasted bad and was tough. Nonetheless, easy meat was going to be eaten by hungry sailors, so the crews of ships that landed on Mauritius slaughtered and ate large numbers of the defenseless dodos.

Much of what is known today about how dodos looked comes from drawings and paintings of an obese and clumsy bird. However, some of the sketches by illustrators may have been of an overfed captive bird. At least one illustration dating back to the 1500s depicts them as slim. Because Mauritius has marked dry and wet seasons, it is possible that dodos fattened themselves on ripe fruits during the wet season and lived on fat reserves during the dry season, when food was scarce. Dodos also may have accumulated body fat for breeding activities and then lost that fat over time.

The dodo was certainly a large bird. It had a huge bill, gaping nostrils, and bulging eyes. Its tail was a clump of feathers positioned high on its back. Adult dodos probably stood about 1 meter (1 yard) tall and weighed as much as 20 kilograms (44 pounds). They nested on the ground and fed on vegetation. Their large beaks seemed to be adapted for shelling and eating fallen fruits or breaking up the roots of plants. It is possible that the horny covering on the beak was shed during non-breeding periods.

Dodos (or the species that evolved into dodos) had been on Mauritius for several million years and evolved, like so many island birds, in the absence of predators. It was their Eden, with abundant fruits scattered on the ground. Dodos were a classic example

Perhaps the most iconic extinct animal is the long-gone dodo, which was a flightless bird from the island of Mauritius off the East African coast. Dodos were very tame and easy prey for sailors. One by one they were killed until the day that the last dodo, probably a bird that had learned to avoid humans, succumbed to old age, leaving behind a poorer universe.

of animal adaptation to remote island life. The sailors reported that dodos had no fear of people, and this trait was part of their undoing. It is likely, however, that many factors contributed to the demise of the dodos, in addition to hunting by humans. There were also habitat change caused by introduced species such as goats, predation on eggs by introduced species (rats, pigs, and crab-eating macaques), and perhaps even natural disasters such as cyclones and floods—disasters they might have well survived in their formerly pristine environment. To put it another way, many of the factors that threaten birds today probably ganged up on dodos and spelled their doom.

The last confirmed sighting of a live dodo was in 1662, less than a century after the bird's discovery by Europeans. Today, we don't even have a taxidermy specimen to study. The last stuffed dodo, stored in Oxford University's museum, was destroyed around 1755 because of its deteriorating state. Only a partial skeletal foot and the head were saved from that specimen, and these fragments include the only soft tissue remains of the species.

An endemic tree in Mauritius, Tambalacoque, is known as the dodo tree. A long-lived species, it has highly valued timber and fruit resembling a peach. In 1973 it was on the verge of extinction, with only about 13 individuals left, each estimated to be about 300 years old. In 1977 it was hypothesized that this tree would disappear because its seeds could germinate only after passing through the digestive tract of a dodo. If so, the dodo and its tree would have been a remarkable example of coevolution. But there is now evidence of germination by the tree long after dodos were exterminated and the possibility of effective seed dispersal by existing species such as fruit bats and parrots. Still, one cannot help but look at the dodo tree and wonder what the scene would have been like with dodos sitting in its shade.

Waiting for the Curlews

For thousands of years the first human inhabitants of Alaska patiently waited for the short nights and long days of June. Summer heralded the arrival of hundreds of thousands of migrating Eskimo curlews, which were hunted for their tender meat. The fragile-looking birds were coming back to Alaska after accomplishing one of the world's most formidable migrations, a round trip they undertook every year between the Arctic and southern South America. Suddenly, in the late nineteenth century, the vast flocks disappeared. The hunters in Alaska still waited every summer, in vain, and eventually even the memories of the curlew arrivals faded like those of long-dead relatives. What happened? How did abundance become scarcity? It turns out that market hunters in the Lower 48 states killed the birds by the trainload to feed America's growing cities. For about twenty years between 1870 and 1890, the curlews flying north and heading south encountered a hail of gunshots.

In his fine 1954 book, *Last of the Curlews,* Fred Bodsworth movingly described in novel form the plight of the birds. He detailed the many natural dangers that the last Eskimo curlews faced in their long migration and how, even after overcoming these perils, many surviving birds ultimately succumbed to shotguns in the U.S. Great Plains. Bodsworth wrote: "But the great flocks no longer come. Even the memory of them is gone... For the Eskimo curlew, originally one of the continent's most abundant game birds... lingers on precariously at extinction's lip. The odd survivor still flies the long and perilous migration from the wintering grounds of Argentine's Patagonia, to... the sodden tundra plains that slope to the Arctic sea. But the Arctic is vast. Usually they seek in vain. The last of a dying race, they now fly alone." The Eskimo curlew is almost surely extinct now, for none has been seen for many years.

No Safety in Numbers

Two hundred years ago, the passenger pigeon was the most common bird in North America, amounting to perhaps 5 billion individuals—roughly as many as all the birds that now overwinter in the United States. The pioneering ornithologist and bird artist John James Audubon observed a passing flock of passenger pigeons for 3 days and guessed that at times more than 300 million were passing per hour and the size of that one flock was perhaps 2 billion birds.

Passenger pigeons consumed huge quantities of seeds from oak, beech, and chestnut trees and formed

The passenger pigeon is perhaps the best-known symbol of extinction. John James Audubon, the first great wildlife artist, captured these graceful birds well in this hand-colored plate.

dense nesting colonies that could stretch for 23 kilometers (14 miles). The sheer weight of roosting and breeding birds was said to have broken branches off large trees and toppled small ones. How do 5 billion birds disappear? The extermination began on America's East Coast, largely through clearing of the forests. Then, after the American Civil War, railroads pushed into the pigeons' midwestern stronghold. Telegraph lines were installed, and soon professional market hunters had a method for being informed of the locations of breeding colonies. Technology, growing urban appetites for meat, and plenty of professional hunters soon resulted in a relentless slaughter. The succulent squabs (young pigeons) were shipped to markets in America's eastern cities, while adults were captured and used as targets in shooting galleries. In 1878 one hunter alone shipped 3 million pigeons from Michigan to eastern markets. An article in *Forest and Stream* magazine described the action in Pennsylvania in 1886: "When the birds appear all the male inhabitants of the neighborhood leave their customary occupation as farmers, bark-peelers, oil-scouts, wild-catters, and tavern loafers, and join in the work of capturing and marketing the game. The Pennsylvania law very plainly forbids the destruction of the pigeons on their nesting grounds, but no one pays any attention to the law, and the nesting birds have been killed by thousands and tens of thousands."

The end for these superabundant birds came with astonishing speed. Market hunting ceased when it was no longer profitable, although there were still many thousands of birds living in large stretches of suitable habitat. Nonetheless, the passenger pigeon rapidly declined to extinction. The cause of the final extinction remains a mystery. Perhaps the best guess is that the birds required gigantic colonies to breed successfully. Individually, they were quite vulnerable to predators, and their breeding strategy may have evolved to swamp predators—appearing suddenly and reproducing before local predator populations could increase enough in size to pose a significant threat. As the colonies dwindled, predator saturation probably became impossible, and the birds disappeared down their enemies' gullets.

The final living member of the species, a female named Martha, died in the Cincinnati Zoo in 1914. Martha still soldiers on, stuffed, in the U.S. Smithsonian National Museum of Natural History, and the species is memorialized in one of John James Audubon's most evocative illustrations. But that's not the end of the story. By devouring billions of seeds, smashing trees, and depositing vast quantities of excrement each summer, these birds shaped the ecology of eastern deciduous forests in the United States. It may be that the mobs of voracious pigeons so reduced the numbers of seeds available to the white-footed mouse that, after the pigeons vanished, big mast years (years in which forest trees produced abundant acorns and other nuts) have produced explosions of mouse populations. As noted earlier, the mice are reservoirs of the tick-transmitted Lyme disease, which now plagues human populations in the eastern United States. Could this be the passenger pigeon's revenge?

The Largest Woodpeckers

The demise of the passenger pigeon is probably the most famous of avian extirpations in North America. Another well-known tale of a bird's extinction involves the massive ivory-billed woodpecker. The bird was (some would say might still be) a denizen of the swamp forests of the southeastern United States and old-growth forests of Cuba. Audubon vividly described the bird's habitat in America 150 years ago:

> *Would that I could describe the extent of those deep morasses, overshadowed by millions of gigantic dark cypresses, spreading their sturdy moss-covered branches, as if to admonish intruding man to pause and reflect on the many difficulties which he must encounter, should he persist in venturing farther into their almost inaccessible recesses, extending for miles before him, where he should be interrupted by huge projecting branches, here and there the massy trunk of a fallen and decaying tree, and thousands of creeping and twining plants of numberless species.*

Almost 50 centimeters (20 inches) long from head to tail, this large bird has been thought to be extinct for a long time. Some hold out hope that the ivory-billed woodpecker still survives in remote areas of southeastern Cuba. A few years ago, hopes soared that it still existed in Arkansas when ornithologists from the famous Cornell Laboratory of Ornithology reported that it had been observed. But subsequent searches for the "Lord God Bird" (so called because of the exclamations that often accompanied first views of

It is sad that the only decent photo that exists of an imperial woodpecker (aside from stuffed skins) is this one—an example of this magnificent bird destined for the cooking pot. Although there are occasional reports of still-living imperial or ivory-billed woodpeckers, someone wishing to get an impression of these largest of all woodpeckers must travel to the Nothofagus (beech) forests of southern South America and be lucky enough to spot a Magellanic woodpecker, a similar species about three-quarters as long.

it) and further analysis of the evidence of the sighting have cast doubt on the claim. There is now a general feeling that the bird the researchers saw was a similar, and more common, pileated woodpecker.

Mexico, unfortunately, has a similar tale of loss. In the oak and pine forests of Mexico's Sierra Madre Occidental, the even larger imperial woodpecker once soared. It was 56 centimeters (22 inches) long. As with its ivory-billed cousin, logging sent the imperial woodpecker, which required huge unbroken expanses of forest habitat, on the road to extinction. The Norwegian anthropologist Carl Lumholtz observed the species in the mountains of Chihuahua in 1902. He recalled: "A Mexican named Figueroa appeared one morning bringing us three superb woodpeckers, magnificent examples of the *Campephilus imperialis*, extraordinarily large ones. This splendid bird stands two feet high; its plumage is black and white, and the male sports a red crest on its head that stands out particularly against the snow."

Hunting also helped send the imperial woodpecker on its way, not just for food (nestlings were considered a delicacy) but also because their feathers were valued for earmuffs. One of the last verified sightings was in 1958, when a member of the Tepehuan tribe (once famous for their revolt against the Spaniards) was seen carrying a dead imperial woodpecker. A dentist and birder named W. L. Rheim recorded that the man described the woodpecker as "un gran pedazo de carne"—a big piece of meat.

From Audubon's magnificent print you can gain some idea of the beauty of the Carolina parakeet, which once graced the eastern half of the United States. And you might sense the great loss we all suffered when what Audubon called the Carolina parrot—the only parrot in the eastern part of the country—was exterminated by human actions.

America's Parakeet

The losses of the world's biggest woodpeckers, like that of the passenger pigeon, were sad events, but there have been several other avian disappearances from North America. One of the saddest was the loss of the colorful Carolina parakeet, which died out at about the same time as the passenger pigeon. The parakeets were widely hunted by farmers, who objected to the large flocks raiding the grain in their fields and eating the unripe fruit in their orchards. The parakeets' habit of hovering in flocks around injured companions made them easy targets.

Other factors hastened their doom, such as capture for the pet trade, loss of habitat (especially old trees with nest holes), and competition for those holes from imported honeybees. It may be that, like the passenger pigeons, the parakeets needed to live in large flocks for security from predators. Ironically, the last Carolina parakeet died in the Cincinnati Zoo in 1914, the same year and place as Martha, the last of the passenger pigeons.

The Paradise and Night Parrots

Members of the parrot family may be among the more vulnerable groups of birds. The paradise parrot in Australia was a beautiful bird that lived in open eucalyptus forests with grassy understories. Their distribution was limited to Queensland and northern New South Wales. These parrots were considered very beautiful, with bright red shoulder patches, a turquoise rump, and a long, bronze-green tail. They were diurnal (active in the daytime) and apparently spent considerable time on the ground feeding on grasses and other plants. Although there is relatively little information on their biology, they are known to have nested in termite mounds. Unfortunately, like others of its kind such as the Carolina parakeet and the Cuban macaw, the species is now most probably extinct. Fires, overgrazing, and land-clearing are some of the causes of their decline, although introduced predators likely were partly responsible for their demise. The last birds were seen in 1927, a long period for a conspicuous species not to be seen if it still exists! But maybe there is hope. The Australian night parrot, a nocturnal (active at night) species of the arid outback, which was feared to be extinct, has just been rediscovered, and excellent pictures have been taken of a living individual.

A Museum Tale

The great auk, as its name implies, was a large bird, standing nearly 1 meter (3 feet) tall. It was flightless—in a way, the Northern Hemisphere's version of a penguin. Totally marine by nature, the species was once extremely abundant and widely distributed in the North Atlantic, from Canada and the United States to northern Europe. Great auks were used as a source of food and as fishing bait by European explorers. The birds' eggs and fluffy down were highly prized in Europe, a fatal attraction that launched the auks' slide toward extinction. Judging from contemporary reports, the auks were very easy to kill. In a single year, tens of thousands were harvested, and the populations in Europe were completely wiped out as early as the mid-sixteenth century. As European exploration and exploitation of the world spread, the North American auk populations began to fall decade by decade. By the 1800s, the great auks had become very rare.

Then a natural catastrophe further spurred its downfall. In 1830 an auk breeding site on an island off Iceland was destroyed in a volcanic eruption. The eruption hastened the end, but the great auk was already fading due to the relentless hunting by humans. Naturalists seeking eggs and skins for European museums and private collectors destroyed the last surviving group of auks in the 1840s. Perhaps the last of its kind, a single bird was reportedly spotted in 1852. No credible sighting occurred thereafter.

The Plight of the Moas

New Zealand is one of the few large island groups that remained free of humans for a long time. Although the whole world had been largely colonized by various human groups over the past 50,000 years, New Zealand seems to have been first settled only a little more than 1,000 years ago. The people came not from Australia, some 2,250 kilometers (1,400 miles) away, but from Polynesia, and were followed a few hundred years ago by Europeans. Each of these human invasions has had profound impacts on New Zealand's biodiversity.

The Polynesian ancestors spawned the distinctive Maori culture of New Zealand, a culture that probably did not interact much with the rest of Polynesia,

Although the Celtic name for the great auk is the source of the term "penguin," and despite their resemblance (made clear in this beautiful print by John James Audubon), this giant extinct seabird is not related to the penguins. Rather, it is related to the razorbill, its closest living relative, and other Northern Hemisphere auks in an entirely different avian order.

due to the great distance between the other islands and New Zealand. It seems likely that the original colonizers brought with them dogs and staple vegetables. Polynesians almost certainly deliberately introduced Pacific rats (kiore) to New Zealand, as they were considered a delicacy and were carried island to island in cages fashioned from hollow segments of the stems of giant bamboos. The kiore were later joined by common (black) and Norway rats, species that commonly infested European sailing ships.

Within five hundred years after the arrival of the Polynesians, half of New Zealand's terrestrial vertebrates were obliterated, including eleven species of flightless birds. Perhaps the most remarkable loss was of the nine species of moas, which were gigantic in size. Moas evolved from a line of flying ancestors that somehow reached New Zealand about 80 million years ago. Living in isolation with no major predators before humans arrived, these birds eventually lost all ability to fly. They fed on plants and, like many herbivorous birds, ingested small stones to grind the plant material in their gizzards. Males and females differed in shape and size, and research indicates that they built shallow nests scratched in the ground (often in rock shelters) and lined with clipped vegetation. We know little of their breeding biology, but there are about forty intact moa eggs in various museums.

Initially it was believed that drastic climate change

Moas were birds of gigantic size. Human invaders made surprisingly short work of exterminating the moa species that once were abundant on New Zealand. They have been gone for only five hundred years or so, and it is still possible to see common plants that had evolved defenses to protect against the browsing of moas.

was the probable cause of the moas' disappearance, but the fact that the remains of all of the moa species have been found in early Polynesian sites strongly suggests that human hunting was the chief factor in their demise. And quite a demise it was. Approximately 160,000 individual moas are thought to have lived in New Zealand before the Polynesians arrived. Evidence from the Shag River Mouth Polynesian site suggests that Polynesians were eating several tons of moa meat annually. There were signs that moa hunters shifted from larger to smaller species as the bigger birds became scarce. This common pattern—killing the big, easy prey first, then shifting to smaller animals—is a classic feature of human overexploitation. It also occurred for shellfish exploited thousands of years ago along the South African coast and is currently playing out in ocean fisheries.

Several factors spelled doom for moas at the hands of humans. One was their low reproductive output. Our best evidence indicates that each female laid only one or two eggs a year, and chicks may have taken five or more years to reach sexual maturity (short by human standards, but quite long for birds). One mathematical model suggests that moas may have gone extinct within one hundred years after the Polynesian colonization of New Zealand. Another model puts the time as somewhat longer: if the first human colonists numbered only 100, and they killed one moa female per 20 people every week, the model estimates the moas would have gone extinct in 160 years. The extinction would have occurred faster if the human population had grown at 2 to 3 percent annually, or if the number of initial colonists were higher, or if the people also ate moa eggs and destroyed their habitats. Moreover, dogs brought in by the settlers quite possibly preyed on moa eggs, thus accelerating their decline. All these factors make the hundred-year estimate plausible. In any case, these fascinating birds were rapidly extinguished.

The only known predator of moas before human arrival was the giant Haast's eagle, which is thought to have been about 40 percent larger than today's golden eagles and able to fly as fast as 80 km/h (48 mph). Female Haast's eagles weighed 10 to 15 kilograms (22 to 33 pounds). Many modern raptors exhibit this reversed sexual size dimorphism with larger females, possibly, among other reasons, allowing males and females to eat different prey species and thus not compete with each other for food. While it is unclear what proportion of each moa species fell prey to the eagle, it is thought that the disappearance of the nine moa species was a major contributor to the extinction of the Haast's eagle. Certainly, habitat destruction and hunting by the newly arrived Polynesians were also factors; indeed, these giant eagles, accustomed to feeding on large bipedal moas, may have occasionally taken children and thus became a target of revenge hunters.

One can imagine what New Zealand looked like before the human invasions, with its towering mountains and rugged coastline—scenes made famous in movies such as *Lord of the Rings*. How amazing it would be to visit the country today if one could see the moas running about and the Haast's eagles circling above! Yet even if the Polynesians had never arrived, it is doubtful that the moas would have survived. The European colonists came long before any conservation ethic took root, and we can assume that they would have seen the moas as easy dinners. But visitors to New Zealand can still see a living trace of the moas' presence—plant species that continue to carry clear evolutionary adaptations designed to lessen the impact of browsing by the long-vanished, tall, plant-eating birds.

The Giant Elephant Birds

Another island that has a somewhat similar geographic situation to New Zealand is Madagascar. It seems to have been first settled by humans around 300 BC. The Malagasy people probably had multiethnic origins and contacts with Arabs. They apparently resisted European invasion until the 1500s. Until that time, it seems that many interesting bird species survived. One example of the effect of European settlement is the fate of the elephant birds. After surviving for 60 million years on Madagascar, the six to twelve species of elephant birds, quite likely the world's biggest birds at the time, vanished in the late 1600s. The largest elephant bird species was 3 meters (9.5 feet) tall and weighed about 450 kilograms (1,000 pounds). Like the dodos, elephant birds evolved flightlessness and large size in the absence of predators. Understandably, the eggs of elephant birds were huge, with a circumference of up to 1 meter (3 feet) and a length up to 34 centimeters (14 inches). DNA has recently been obtained from the fossilized eggs of elephant birds,

Until four hundred years ago or so elephant birds, heavier than the biggest moas, roamed Madagascar. As was noted, one of their huge eggs could supply omelets to an entire family group, so it is no wonder that invading modern humans quickly wiped them out.

and there have been suggestions that they might be cloned in the future (don't hold your breath; it is much harder to clone an extinct species than we once thought it might be).

The elephant birds were related to ostriches and emus—birds we still have today, luckily. They were powerfully built, with long necks, spear-like beaks, and massive legs with sharp claws. Unlike ostriches, however, they probably were not swift runners. Their feathers, like those of emus, were bristly and hair-like. In spite of their fearsome appearance, elephant birds ate only plants.

It is not clear whether elephant birds were exclusively found in the rain forest of Madagascar, but only elephant birds dispersed seeds of some rain forest palm species. Similarly, fruits of some endemic Malagasy plants, such as those known as cats' claws, had hooks that might have evolved to attach to the feet of elephant birds for dispersal. On the other hand, some other plant species had spines that might have evolved to resist browsing by elephant birds. As with the moas, traces of plant-animal coevolution can probably still be seen in the flora.

The presence of elephant bird eggshells near early hearths suggests that people enjoyed giant omelets. Again, as with their moa relatives, the slow breeding of elephant birds might have made population recovery impossible, especially under the pressure of heavy egg harvesting by humans. The dogs and rats brought by people to Madagascar may also have preyed on elephant bird eggs, accelerating the birds' demise in a manner similar to that of the moas. Other likely factors contributing to the extinction of elephant birds include habitat loss, the transfer of diseases from chickens, and a progressive drying of Madagascar's climate during the past 12,000 years.

Tibbles the Cat

The Stephens Island wren was an odd bird indeed. Tiny, nocturnal, and flightless, it was an insect hunter that lived mainly on the southern of New Zealand's two main islands. The introduction of Norway rats, however, quickly eliminated the bird from all but one offshore islet. By the 1840s the wren lived only on Stephens Island, which measures only 150 hectares (0.6 square mile), about half the size of Manhattan's Central Park.

The wren was erased from Stephens Island in the early 1890s after the construction of a lighthouse. The direct cause was predation by cats, and for many years it was assumed that a single cat, the lighthouse keeper's pet called Tibbles, was responsible. The cat sometimes brought dead wrens to the lighthouse, and the keeper was careful enough to save one. He gave it to one of the engineers, who skinned it and sent it to London as a museum specimen. This was simultaneously the first scientific recognition of the species and the end of it. Much is unclear about events on the island, but it seems likely that one pregnant cat or several cats landed with the lighthouse crews and eventually multiplied and hunted the wrens to extinction. A search in early 1895 found no wrens left. Tibbles and his friends had eradicated the last population of one of the few flightless songbirds the world has ever known.

The Mexican Galapagos

Mexico's Guadalupe Island is remote; lying in the Pacific Ocean 250 kilometers (150 miles) from the Baja California peninsula, it was formed by volcanic activity thousands of years ago. It is now 35 kilometers (21 miles) long and 10 kilometers (6 miles) wide. It was "discovered" by biologists in 1875, although it was already well known to whalers. One of the first naturalists to visit, Edward Palmer, found a wide variety of unique birds as well as other life-forms. Palmer compared his discovery to those made by Charles Darwin from his research on the Galapagos Islands.

Because of its distance from the mainland, species that landed on Guadalupe became isolated from mainland forms and evolved into different species. But, as happened in the Galapagos, people brought problems to Guadalupe. They introduced goats, dogs, cats, rats, and mice, which devastated the biological paradise. Some introductions were deliberate: whalers and fur hunters released some goats in the nineteenth century to generate a supply of fresh meat to be available when their ships visited; both fishermen and whalers intentionally introduced cats and dogs. And all accidentally transported rats and mice to the island.

By the beginning of the twentieth century, twenty-six species of plants and eleven species or subspecies of birds had been driven to extinction on Guadalupe Island. Things got even worse as the goat population soared. By 1956 an estimated 60,000 goats were on the island, eating everything they could stomach. The island became a wasteland. The once extensive forests of junipers, oaks, and pines, which had been covered by epiphytes (plants that grow on other plants), had virtually disappeared. Today, only a small remnant forest remains in the north of the island, and all the trees in that forest are very old.

Cats and rats eliminated the tiny Guadalupe Island petrel, which nested in cavities in the ground and had no defenses against predators. Sadly, the island's unique caracara was exterminated by poisoning and shooting by goat herders, who considered it a danger to the goats. In a situation similar to the tragedy of the great auk, a scientific collector killed the last eleven caracaras in 1911 and sent their remains to a private collection in England.

Despite such losses, the Guadalupe biodiversity that is left still has a chance. Recently, the introduced goats, cats, and dogs were eradicated, and the native vegetation is making a remarkable comeback. Most species of flowering plants, thought to be extinct for more than a century, have reappeared, and populations of the surviving species of birds are rebounding. The island is really a model for what might be done in many other, similarly devastated places. What is gone is gone, sadly enough, but perhaps our generation can be judged by what we did to save and restore the remaining flora and fauna.

More than a Vacation Destination

The Hawaiian Islands were "discovered" by James Cook in 1778, long after the Polynesians had established populations on those volcanic outposts. The earliest human inhabitants valued the gorgeous feathers of many bird species and hunted them to make capes and other decorations. But Cook's arrival heralded an age of habitat clearing; releases of cattle, goats, and rabbits; and the introduction of a mosquito that carries avian malaria, avian poxvirus, and a protozoan parasite named *Toxoplasma gondii*. These European introductions added to the problems created by the much earlier Polynesian introductions of rats and pigs.

Malaria proved especially devastating, extirpating many Hawaiian bird populations at lower altitudes where the mosquitoes thrived. Avian malaria actually seems to have caused the serious decline or extinction of as many as sixty endemic forest bird species on the islands. Without having evolved in the presence of avian malaria, Hawaiian bird populations were simply unable to cope with the novel disease. A bit of good news is that some of the species that survived the establishment of the disease, such as thrushes, are now showing resistance, although numerous other species, such as the endangered Hawaii creeper, remain vulnerable.

In 2002 the Hawaiian crow, or alala, was added to the list of species that are extinct in the wild. It was historically found on the "big island" of Hawaii, but bones discovered on Maui indicate that it may also have occurred there before human contact. While the crow was initially found in both dry and moist forests and at elevations as low as 300 meters (960 feet), by 1992 the 11 or 12 remaining wild individuals of the species were confined to montane areas. That was the last known year in which a wild chick fledged.

Inhabitants of continents often think crow species are pests (as they can be), but the Hawaiian crow, or alala, is a different story. It is now extinct in the wild, extirpated by the same set of threats that have punished much of Hawaii's fascinating avifauna: diseases introduced with mosquitoes, deforestation, predation by introduced mongooses and rats, and destruction of native vegetation by introduced pigs. Ironically, attempts to reintroduce individual alalas to the wild have suffered from depredations by the Hawaiian hawk, itself endangered.

The Hawaiian crow's decline was due to familiar causes: predation by introduced species, especially rats and Asian mongooses; exposure to diseases; and habitat loss. Feral cats spread toxoplasmosis, a disease that can afflict the crows. Coffee farmers and ranchers who considered them agricultural pests also persecuted the crows. Farmers would attract Hawaiian crows, by imitating their calls, and then shoot them. Saving the species meant that it could no longer be part of the wild. Today more than fifty individual Hawaiian crows survive in various breeding centers, but there is concern that the captive crows may suffer from inbreeding. Reintroduction into the wild remains the only possibility for resurrecting this species, and in 2009 the U.S. Fish and Wildlife Service announced a five-year plan, costing $14 million, to prevent the extinction of this species by protecting habitat and ameliorating threats (i.e., removing introduced predators). It will be a difficult project; similar efforts for other species have sometimes worked, sometimes not. But it is worth trying, on the chance that the species will eventually recover.

Two scarlet macaws wait on a branch before landing at a colpa (clay-lick) in the Amazon. The parrots in western parts of the forest frequent bare riverbanks and, especially when breeding, eat clay. Eating clay was once thought to help neutralize toxins in the seeds the birds dine on, but recent research suggests this is wrong; most likely the clay is ingested for its salt content, since sodium chloride is scarce far from the oceanic winds of the eastern coast.

5 BIRDS IN TROUBLE

America's conservationist president, Theodore Roosevelt, knew what it meant to lose a species of bird. He once famously stated,

> *The extermination of the passenger pigeon meant that mankind was just so much poorer; exactly as in the case of the destruction of the cathedral at Rheims. And to lose the chance to see frigate-birds soaring in circles above the storm, or a file of pelicans winging their way homeward across the crimson afterglow of the sunset, or a myriad terns flashing in the bright light of midday as they hover in a shifting maze above the beach—why, the loss is like the loss of a gallery of the masterpieces of the artists of old time.*

Unfortunately, Roosevelt's fears became realities all too often as the twentieth century unfolded. Today thousands of bird populations and many entire species hang on by their wingtips. Large portions of our avifauna seem destined to disappear in this century if we continue with business as usual, as increasing climate disruption adds to the effects of habitat destruction, hunting, and other assaults humans have imposed on the world's fauna. Beyond the species themselves, we will also lose the important roles birds play in our lives, from giving us esthetic pleasure to eating various insects that plague people. This chapter describes some of the birds we have lost in more recent times or are now close to losing.

Midnight for the Macaws

The last known free-living individual of Spix's macaw was seen in Brazil in the year 2000. The Spix's macaw got its common name from its discoverer, a German naturalist named Johann Baptist von Spix. Overharvesting has been the main culprit in the demise of this species. But people generally didn't eat Spix's macaws; they captured them from the wild to be sold as house pets because of their great beauty and rarity. Even when the macaws were on the brink of becoming extinct, many reports indicate that capturing and sales continued until none were left to be trapped, caged, and sold. The species is now presumed to be extinct in the wild.

Genetic evidence supports the view that, even before the wild birds were turned into pets, Spix's macaws survived in nature in extremely small numbers. They were predominantly found in woodlands along watercourses in otherwise treeless landscapes (so-called gallery forests) that were dominated by caraiba (trumpet) trees in which the birds nested. They usually fed on the fruits of two plant species belonging to the family Euphorbiaceae. The demise of the wild Spix's macaws at the hands of the pet trade was accelerated by the loss of the forests and perhaps by the shooting of wild birds.

Another possible human-fostered problem for the macaws was the invasion of their native habitats by aggressive Africanized honeybees. Honeybees are native to Africa, where they have long been subject to nest-raiding by mammals and birds for their honey and wax. People domesticated bees centuries ago, producing a relatively calm "Italian" strain, which has been introduced over much of the world, including in Brazil, where it was not a great producer of honey.

A well-known scientist, Warwick Kerr, thought that by hybridizing African and Brazilian bees he might be able to create a strain that was both calm and yielded a rich lode of honey. But before Kerr could do the experiment, a Brazilian beekeeper visiting Kerr's lab deliberately allowed some of the African bees to escape. That beekeeper's motives are unknown, but besides possibly hastening the exit of Spix's macaws, he has been responsible for the deaths of numerous human beings.

In a poignant twist, the last known wild male Spix's macaw was discovered in 1990 paired with a female, but the female was not of his own species. The male Spix was trying to reproduce with a female blue-winged macaw. The mismatched couple did mate, and she even laid eggs, but, as one might expect, the eggs were infertile. Conservationists tried to get the male hitched to a female Spix's macaw by releasing a formerly captive female Spix's into the territory of the interspecies pair, but the male showed no interest in her. Unhappily, she collided with a power line and disappeared, presumably dying from her injuries. This episode is an example of the difficulty of acclimatizing captive birds and successfully re-establishing species in nature. The "odd couple" presumably remained paired until January 2000, after which they too disappeared.

More than seventy Spix's macaws now live in captive breeding programs run by conservationists. To counter the risk of losing genetic variability due to inbreeding, individuals have been exchanged between various institutions in an effort to maintain their genetic diversity. But conservationists are afraid to release the macaws, worried that the birds might end up being trapped and sold as pets. Several captive macaws are currently held by the Al-Wabra Wildlife Preserve (AWWP), far from their native Brazil in the Middle Eastern country of Qatar. In 2009 AWWP announced that it had purchased 2,200 hectares (5,437 acres) of land in Brazil at the site where the last wild Spix's macaw had been observed. In a careful, step-by-step process, domestic livestock are first being removed from the site to facilitate habitat recovery, especially regeneration of the trees essential to the birds' breeding. This commendable effort demonstrates how much money and effort are required to give a single bird species a chance and how much easier it would have been simply to protect the wild populations that once existed.

Lonesome as the last wild Spix's macaw was, its plight has been shared by virtually all the seventeen other species of macaws. These gorgeous, long-lived, highly social big parrots are all threatened by deforestation and capture for the pet trade. Hyacinth macaws, the giants of the group, are down to a few thousand individuals. Their close relative, Lear's macaw, is now represented in the wild by only one hundred or so birds in the interior of the northeast bulge of Brazil, where the stands of palm trees on which they depend for food have been reduced to tiny remnants. Fortunately, though, there is great interest in both species, and local conservation groups and landowners are working to keep them wild.

Is it really so important to have a bird in one's apartment in New York or Manaus or Hong Kong that we as a society are willing to permit unregulated, or barely regulated, trade in avifauna? The problems that vex macaws are human problems. We seem to value someone's right to have a pet, or to own a patch of land, or to harvest timber over our ethical and self-interested responsibility to protect nature. Then, when a species reaches a crisis level, we sometimes react, spending ten or a hundred times what we would have needed to spend had we just strongly enforced a policy that simply said no, stop, leave this species in peace. Even better, if we could just adopt a global policy of humanely and fairly limiting the scale of the human enterprise, gradually reducing the population size of *Homo sapiens*, curtailing overconsumption by the rich (while increasing needed consumption by the poor), then we might leave some room for the natural systems all humanity depends on.

The Whooping Crane

The whooping crane is the poster child for crane conservation. It is a tall (1.5 meters, or 5 feet), beautiful white North American bird that has teetered on the edge of extinction for decades. Its numbers once were in the tens of thousands, but at one point as few as fifteen to twenty individuals remained. The destruction of wetlands was the main cause of their decline, though many were also shot. In the United States and Canada government agencies worked intensively and creatively to reverse the decline. Today there are several hundred birds in the wild and about 150 in captivity. Even so, the species is hardly safe.

In 2013 the U.S. federal government came into conflict with the government of Texas, a state suffering a prolonged (and possibly climate change–related) drought. Because of the drought, insufficient water was reaching coastal San Antonio Bay, the wintering grounds of one of the main whooping crane populations. This resulted in the deaths of twenty-nine cranes. A federal judge ordered the allocation of water from two key rivers to aid the vital crane habitat. For the wetlands-dependent cranes, the conflict with agricultural uses of water is likely to be a persistent battle. Rain may eventually end the conflict, but with ongoing climate disruption drought may become more frequent. The cranes had a narrow escape this time, but all too often, short-term economics trumps the ecological needs of wildlife.

Endangered Eagles

One large bird that has been successfully rescued from extinction is the bald eagle, the avian symbol of the United States. This famous raptor was once approaching extinction, declining to fewer than five hundred breeding pairs in the middle of the twentieth century but recovering to more than ten thousand pairs early in the twenty-first century. Eagles suffered from habitat clearance, reduction of food sources, hunting and random killing, and perhaps most important, from DDT (dichlorodiphenyltrichloroethane) poisoning. The accumulation of DDT in female eagles led to thinning of eggshells, and the parents ended up sitting on omelets instead of eggs. The bald eagle was saved when it was listed as endangered under the U.S. Endangered Species Act (1973), which banned hunting

The whooping crane, an impressive bird with a wingspan of up to 2.4 meters (8 feet), is now threatened by climate change, which probably contributes to more frequent droughts in its wetlands breeding grounds.

of eagles and initiated efforts to protect their habitats. Similar efforts in Canada and the banning of DDT in North America ultimately led to the strong recovery of the species.

Several large tropical eagles, however, haven't been so lucky. The tropical American harpy eagle has a wingspan of up to 2 meters (6.5 feet) and is capable of snatching monkeys out of jungle treetops. Flying at up to 80 km/h (48 mph) and armed with talons the size of grizzly bear claws, they are formidable predators. Harpy eagles prey on monkeys, squirrels, coatis, sloths, porcupines, and a wide variety of other animals. Formidable as they are, they are declining in the face of the activities of that other formidable animal, *Homo sapiens*. Harpies require large tracts of unbroken forests as hunting grounds and cannot survive the expanding footprint of civilization as throngs of people busily clear the lands and create human-dominated landscapes throughout Central America. Once-vast expanses of impenetrable jungle habitats are quickly being reduced to scattered forest fragments.

To make matters worse, this ferocious predator has no protection against the darts, arrows, and bullets of human hunters. Harpies do occasionally attack domestic animals and are rumored (almost certainly falsely) to sometimes take human babies. These

Tropical raptors such as the ornate hawk-eagle have been decimated by habitat fragmentation, declining populations of prey, and hunting. Eagles are the top aerial predators in the tropics; they will survive only if enough habitat is protected.

stories encourage the hunting of these so-called pests. Adding to their problems is that the species is a slow breeder; a pair usually produces a chick only every couple of years. As a result, the birds are virtually extinct in Central America and are rapidly declining in South America.

The Philippine monkey-eating eagle is slightly longer than the harpy and is similar in behavior and requirements. It faces essentially the same array of threats as its New World cousin. But because of its much more restricted geographic range, it is more critically endangered than the harpy eagle. There are perhaps fewer than five hundred Philippine monkey-eating eagles left in the wild. Human population growth in the vastly overpopulated Philippines (already twice the population density of China and ten times that of the United States) threatens this wonderful eagle, as well as the rest of the avifauna and, ultimately, the Filipinos themselves.

Some of the blame for their disastrous situation goes to the late, and perhaps aptly named, Cardinal Jaime L. Sin (1928–2005), who waged a relentless war on women and birth control, and therefore indirectly on biodiversity. The rapidly growing human population has spread out, clearing land, destroying natural ecosystems, and occupying hazardous marginal areas, often with inadequate infrastructure (e.g., shanty towns). The disastrous effects of Typhoon Haiyan in 2013 were in part a result of poverty and overpopulation.

Aerial Plankton

One worrisome feature of the general decline of birds in the face of expanding human activities is the severe reduction of the species that capture insects on the wing. The sky is replete with aerial plankton—insects and other small arthropods such as ballooning spiderlings, many millions of them found in each square kilometer (0.6 square mile) of sky from housetop height to 150 meters (500 feet). They are fed on by birds such as swallows and swifts that stay aloft for major portions of their lives, unlike flycatchers, which perch in high places and only sally briefly after prey.

A recent report from Connecticut Audubon Society scientists noted that, in the United States, chimney swifts, purple martins, and common nighthawks, or "bullbats," have become ever more rare and local. The reasons are unclear and could range from the replacement of favored gravel-rooftop egg-laying sites with modern roofs (for nighthawks) and loss of human-supplied nesting sites (for martins) to DDT acquired on overwintering grounds in South America or diminished concentrations or changes in composition of the aerial plankton. The reasons for the last, if they are occurring, are also unclear.

Declines of these winged insect vacuum cleaners appear to be widespread, including, for example, a dramatic reduction of whip-poor-wills in Canada, but data are mostly lacking. Other aerial insectivores (insect-eaters) such as barn swallows (the most widespread of them) and bank swallows also appear to be in trouble. What seems certain is that the aerial insectivores are major predators on mosquitoes, which are not just annoying pests, but are also important vectors in North America of West Nile virus and viral encephalitis, as well as dengue fever, yellow fever, and malaria elsewhere. Thus the woes of these graceful birds might translate into woes for us.

Walking Birds

Among the exceptions to the disappearance of flightless birds is the 114-kilogram (250-pound) ostrich, the biggest living bird. Ostriches are one of the glorious features of African ecotourism, with their spectacular feather displays and smooth running (up to 70 km/h, or 45 mph). If cornered, they can be deadly, striking out with their extremely powerful legs and sharp talons. When threatened, they usually try to outrun the danger, but sometimes they seek concealment by stretching their long necks along the ground, leaving only the humps of their bodies visible. This habit apparently led to the myth that, like many politicians, they "hide their heads in the sand."

Ostrich numbers have gone way down in the past few hundred years. The species has become extinct in the Near East and is critically endangered in North Africa. It is scarce in the wild in many parts of sub-Saharan Africa. Nonetheless, it is still widespread and protected in reserves and on ostrich farms, so for the moment they endure. How long that will last with the

Emblematic of Africa, the ostrich is both the largest living avian and the fastest runner. It has declined everywhere but is still widespread, although its Middle Eastern population went extinct a half-century ago. It is now farmed for meat, and one occasionally gets its revenge by disemboweling a person who foolishly corners it.

booming human population and a potentially huge increase in hungry people in its sub-Saharan stronghold remains to be seen.

The ostrich is a ratite, a member of a group of bird families that some scientists think descended from a species of flightless ancestors that lived on the southern supercontinent Gondwana. The separated populations of the ancestral species evolved and diverged when Gondwana broke up through plate tectonics (continental drift), beginning some 180 million years ago. Australia's largest bird, the emu, is also a ratite. It weighs in at 18 kilograms (40 pounds). Emu numbers seem high enough and their populations stable enough that one might describe them as "secure," although the subspecies on Tasmania was wiped out, and emu distribution elsewhere has clearly become restricted to certain areas. Emus do well in arid areas and can be raised for meat—important insurance policies for this drought-resistant species, which occupies an overpopulated, water-short continent that may well become even drier in the near future.

The other Australian ratite, the fruit-eating southern cassowary, is fading away because of habitat destruction, attacks by feral dogs, and nest predation by feral pigs. The cassowary is big (60 kilograms, or 132 pounds) and can have a nasty disposition; there are cases of the bird attacking people who provoke it. It sometimes hangs around campgrounds seeking food, and lucky campers may see a big male guarding a cute striped chick. But prudence says not to go too close; the bird's powerful legs ending in sharp-nailed feet can be lethal. The southern cassowary seems doomed to join the many other creatures that have already gone extinct on the biologically diverse but highly vulnerable continent of Australia.

The ratites are represented in South America by the flightless rhea. At about 23 kilograms (50 pounds), rheas are only about one-fifth as big as ostriches. But like the cassowaries, they are fading. Despite their wide distribution, they are being crushed by rapid conversion of their grassland habitats to agriculture, people killing them as agricultural pests, and hunting to procure their skins and meat. In an odd twist, a band of rheas escaped from a farm in Germany and established a small feral population. This population is now protected and living in Germany much as they would in South America.

The famous kiwis of New Zealand are also flightless, and arguably the strangest of the ratites. There are five species of the strong-legged, nocturnal kiwis. They all have very long bills tipped by sensitive nostrils that help the birds find invertebrates and fruits in the soil. Their feathers have evolved into fascinating hair-like structures. Kiwis usually mate for life, and females lay eggs that are, in proportion to their body weight, the largest of any bird's. They are slow breeders—one egg a year—with the chicks hatching after ten to twelve weeks of incubation. Once hatched the chicks spend very little time in the nest, at most a few days, before they begin to feed themselves.

All the kiwi species are threatened with extinction. The problems they face range from imported predators (especially dogs, cats, and stoats), to habitat destruction and being run over by cars. The two species of kiwi most in peril are the rowi and the shy Haast population of the tokoeka. These particular kiwis both live in the cold, mountainous South Westland. Their numbers are estimated to be in the hundreds, and probably the greatest problem they face is the eating of chicks by introduced stoats, also called short-tailed weasels. All unmanaged kiwi populations

All species of New Zealand's kiwis (*top*) that are not being managed are slipping away. These flightless, nocturnal birds are quite vulnerable to introduced predators, including rats, cats, opossums, and dogs. The famous and very endangered kakapo (*bottom*), the flightless parrot of New Zealand, is still nocturnal long after the extinction of its daytime predator, Haast's eagle, which plagued its ancestors and the moas.

are in decline, but the good news is that community groups and foundations trying to protect them are becoming more common. It is possible that the efforts of these groups, combined with those of the New Zealand Department of Conservation, may improve the conservation situation for kiwis.

One of the rarest and most unusual flightless birds on Earth is the New Zealand kakapo. It is the only flightless parrot, and also the heaviest. Males can weigh more than 2 kilograms (4.4 pounds). They can store large amounts of body fat and have a strong odor and an amazing, booming mating call. Kakapos originally were a prey item of the now-extinct Haast's eagle that once dined on moas; in response to that threat, they evolved into a nocturnally active species that possesses a camouflage-and-freeze daytime defense.

The kakapos' defenses proved virtually useless, however, against the depredations of the scent-hunting ferrets, stoats, rats, and cats that people introduced. This predation, combined with habitat destruction, reduced the birds to fewer than one hundred individuals. In an emergency effort to save the species, the remaining wild kakapos were gathered about two decades ago by conservationists and transferred to several predator-cleaned small offshore islands. There are now about 120 of these slow-breeding birds.

Swimming Birds

Less unusual looking than the kakapo or the kiwis is the flightless cormorant of the Galapagos Islands. It's the largest species in its family. The flightless cormorant has always had a very restricted range, but makes its living by underwater fish-hunting, the same as other cormorants. It lost its power to fly under a virtual absence of predators. But unlike the kakapo, which once had a greater range, this cormorant was always restricted to one small island group. Currently, there seem to be fewer than two thousand individuals left, with feral dogs having been their most serious problem. Fortunately, the dogs have been exterminated on one island, but the possibility of rat or cat introductions and oil spills still threaten the cormorant. Also, some cormorants are killed in fishing nets, and human fishing operations compete with cormorants for food. Adding to the list of dangers, climate change, perhaps in the form of more severe El Niño events (climatic patterns triggered by periodic warming of the tropical eastern Pacific Ocean surface and possibly intensified by climate change), may destroy the species in one terrific storm or set of storms, should its population size sink too low.

The last group of flightless birds in trouble is a familiar one: the penguins. We think of this group as Antarctic, but one species, the Galapagos penguin, actually lives near the equator and in the Northern Hemisphere. Not surprisingly, the same forces that imperil the flightless cormorant threaten it. The more southerly penguins live around the Antarctic continent and northward on small islands, as well as in New Zealand and on southern coasts of Australia, Africa, and South America. In some places it is relatively easy to wander through colonies of hundreds or thousands of individual Adelie, chinstrap, gentoo, or king penguins. One of the great nature experiences is listening to the chatter of reuniting pairs, watching and hearing adults on return from foraging trips to the sea regurgitate food for impatiently waiting chicks, and observing birds sneakily stealing pebbles from the nests of other couples. Two-legged animals on land, however, are not seen as threats.

Frequently, though, the penguins react defensively as predatory petrels or skuas (seabird relatives of gulls) zoom overhead looking for an opportunity to snatch an egg or a hatchling. At the edge of the ice one can see the penguins go through an "after you, Alphonse" routine, urging others to enter the waters first in order to observe whether they are devoured by the voracious leopard seals often waiting for a meal. Killer whales also hunt them at sea.

The ancient threats of natural predators are certainly worry enough for the penguins. But now they face an exploding fishery in Antarctic waters as human beings try to find ever more protein in the depleted oceans for their burgeoning populations. The fishing vessels drain the penguins' food supply, requiring the birds to spend more time hunting in the dangerous seas. Penguins on South American and African coasts are also often severely impacted by oil pollution, frequently caused by illegal (and immoral) discharges of oil in ballast water by oil tankers.

The biggest threat of all to Antarctic penguins is human-caused climate change. The penguins rely on the distribution of sea ice and krill (small crustaceans) that are critical to the Antarctic food web. Both the amount of sea ice and the abundance of krill are

The star of the movie *March of the Penguins,* the emperor penguin is the largest living penguin. It breeds in winter on the frozen coasts of Antarctica, and the male uniquely balances the egg on his feet for two months to keep it off the frozen surface and under his brood patch until it hatches. Like most polar animals, this penguin is considered to be under threat from climate change.

expected to change rapidly in the next few decades. These changes may well doom some penguin populations.

The Largest North American Bird

Even soaring against the vast backdrop of the Grand Canyon, the California condor immediately seems huge. At one time these great birds rode rising thermal airflows across a large part of North America. This scavenger species once ranged from New England to Canada's British Columbia and from Florida to Mexico's Baja California. Its range and numbers declined dramatically, starting ten thousand years ago when many of the Western Hemisphere's large mammals were wiped out during the Pleistocene overkill. The destruction of those mammals meant the loss of much of the condors' food supply, which consisted of carcasses of creatures such as mammoths and giant ground sloths.

Condors were still locally common until recent times, but their numbers declined rapidly during the twentieth century due to habitat alteration, electrocution by power lines, lead poisoning (from lead shot ingested when scavenging remains of hunted animals), and poaching. Condors are another of the slow-breeding animals that suffered a rise in death rate, a combination that, unchecked, eventually leads to extinction. The population of condors dropped to an astonishing low of twenty-two birds in 1983. A few years later they became extinct in the wild when the last six wild individuals were captured and added to the existing birds held in a coordinated captive breeding and species recovery program. The genetic diversity of the captive stocks was maintained by animal exchanges between different breeding facilities.

Normally, adult female California condors lay one egg every other year, but if the egg breaks or the chick dies, they often lay a replacement egg. Biologists used this feature of the species to aid in its recovery. They removed the first laid eggs from the nests of the captive birds and artificially incubated them. The San Diego and Los Angeles zoos have tirelessly carried out this complex and massive conservation effort. Researchers feared that condors raised by humans might become imprinted and think the humans were their parents. So, in what was rather a comical scene, condor-like puppets fed the hatched chicks. Adult condors were also placed in the pens of puppet-raised young to further alleviate concerns that the hatchlings might develop behavioral misperceptions.

Since 1991 condors have been gradually reintroduced in the wild. To minimize the chances of the captive birds catching diseases when released in the wild, they were immunized against diseases such as West Nile virus. The first release was bumpy, as condors landed on houses, walked along roads and highways, and begged for food from disconcerted humans. Sadly, some released condors died from collisions with power lines or after drinking antifreeze dumped along the highway. These failures led to initiation of human-aversion training of captive birds. For instance, human structures were placed in the pens, and these delivered mild electric shocks on contact so that young condors learned to avoid them in the wild. Similarly, zookeepers deliberately harassed captive condors so they would not approach humans when released.

Today there are more than 350 condors in existence, of which 180 are in the wild. The population and the range of wild birds are increasing, with populations now re-established in California, Arizona, Utah, and Mexico. The release of California condors in the San Pedro Mártir National Park in the Baja California peninsula of Mexico was a triumph of international diplomacy between the two countries. The California condor had disappeared from Mexico in the 1930s. For decades the silhouette of the condor was absent from the deep canyons and majestic mountains of the national park. In 2011 the birds were reintroduced to that remote region, which used to be part of their range, bringing back the condor's shadow to those ancient mountains.

Ingestion of lead from carcasses remains a major threat to wild condors, but there have been programs to educate hunters and promote the use of non-lead ammunition. In 2008 California passed legislation

This giant vulture, the California condor, once dined on the remains of the Pleistocene megafauna. Now the increasing numbers released into the wild must perform their ecosystem sanitation services on smaller creatures.

that bans the use of lead bullets when hunting in the condor's range. In recent years, though, the proliferation of wind turbines, a form of green energy, have unexpectedly threatened the California condors. This example indicates how complex and dynamic conservation issues often are.

Herculean human effort seems to have saved the California condor from extinction, at least for the time being, but conserving this one species has cost more than U.S. $35 million. Whether it would have been better to use that money on other conservation efforts is often discussed, but the point is moot. Charismatic species will attract the most funds for conservation, so if that money had not been spent on condors, there is no assurance that it would have been spent on conservation at all.

The Rare Rail

An effort similar to that lavished on the California condor was made in an attempt to save a much smaller and more obscure bird, the Guam rail. These once common marsh birds, found only on Guam, declined precipitously by the early 1980s, largely due to predation by the brown tree snake, which was accidentally introduced to the island, most likely as a stowaway in a cargo plane from somewhere in Melanesia, after World War II. The few rails that remained in the wild were captured and put into captive breeding programs. Later, in 1998, some of the captive rails were reintroduced on Guam, but by 2002 no rail was detected, even in an area made snake-free. There was evidence of breeding by the released rails, but no living rails could be found.

In a parallel effort begun in 1995, more than one hundred Guam rails were introduced to the nearby snake-free island of Rota. There have been reports of breeding by the released rails on this island, but predation by feral cats seems to be a major problem. It is estimated that only a few of the released birds persist on Rota. This difficult situation is not without hope: more than one hundred Guam rails exist in various captive breeding programs throughout the United States, and there are plans to release more rails on Guam. The plan is basically simple: eradicate the tree snakes and feral cats from potential release areas and hope the population can re-establish itself. How well all these efforts will work remains to be seen. The rest of Guam's native avifauna was not as lucky as the rail. Of eleven species of forest birds found on the island when Europeans first arrived, none had evolved mechanisms to deal with a predator like the brown tree snake. They all vanished soon after the snake's appearance.

On the Caroline Islands to the south of Guam, many populations and species have followed the path of Guam's birds, even in the absence of brown tree snakes. While some species, such as the Micronesian starling, the Caroline Islands reed-warbler, the Caroline Islands swiftlet, and the endemic Pohnpei lorikeet, seem to be doing well, other birds may be on their way out. The Truk white-eye is hanging on primarily in a patch of forest high on the island of Tol in the Chuuk (or Truk) atoll. Its habitat overlooks the famous lagoon that was once the main base of the Imperial Japanese Navy, where the whole sea floor is littered with the coral-encrusted remains of Japanese ships sunk in air raids in 1944. Faring slightly better, for it still lives, if barely, is an all-white passerine called the Truk monarch. This endangered bird, a flycatcher, is one of the most thrilling sights on the island. When perched, it spreads its pure white tail and flicks it from side to side, putting on quite a show.

The general trend in the Caroline Islands, as on Hawaii and on most other islands in temperate and tropical seas, is a one-two punch. The first blow is destruction of habitat, especially forests, and the second is the introduction of deleterious species. Both of these negative impacts can be traced quite easily to the increase and spread of people.

A similar decline can be seen in birds that roam the open seas, for they too depend on islands to complete their lifecycles. The group known as tube-nosed seabirds—petrels, albatrosses, shearwaters, and their relatives—includes some of the most severely threatened bird species. These seabirds are unfamiliar to most people, as even those who have spent time on ships often see them only as distant objects sailing low over the waves. Even though some tube-noses—such as Wilson's storm petrel and the short-tailed shearwater—are still abundant, overall, tube-noses are one of the most endangered bird groups. Several species have globally distributed populations numbering only a few hundred individuals.

We who have been fortunate enough to be on ships in sub-Antarctic waters have seen huge albatrosses at almost arm's length as they soar just off a vessel's bridge. Even more awe-inspiring is watching albatrosses on remote islands come in with food for their chicks, each one alone atop a cylindrical mud nest. Landing on hard ground is a dangerous business for these heavy, slender-winged, turkey-sized birds. One misjudgment could lead to a crash, a broken wing, and, ultimately, death for both parent and offspring. Because of the danger, an albatross may make several passes before finally deciding to land. Like airplane pilots, they attempt to "stall" (increase the angle of their wings to the point where they no longer support the bird's weight) just when they want to touch down. A patient observer can sit close enough to see small feathers lift up as boundary layer separation occurs; the smooth flow of air right next to the wing separates from it, producing turbulence and a loss of lift that causes the bird to stop flying. This process has practical applications. Arrays of feathers are sometimes glued to the wings of aircraft being tested in wind tunnels to study the pattern of boundary layer separation.

Tube-nosed birds generally nest on the ground on predator-free islands. That's the rub—people have introduced rats, cats, and other predators to those islands, much to the distress of the species that breed there. Tube-noses have also been hurt by ocean pollution, including from accidental ingestion of plastic litter, millions of tons of which are dumped into the once pristine oceans each year, killing thousands of fish and marine mammals and birds. There is some sign that Laysan albatrosses purposely feed on plastic debris, which they regurgitate for their chicks, often killing them. In their nesting grounds on islands such as Midway, the locations of dead chicks are marked by little piles of immortal plastic surrounded by rings of decaying bones and feathers. Equaling and sometimes exceeding these problems are the destructive practices of large-scale commercial fisheries. Humanity's growing appetite for seafood means that many of the tube-nosed species, especially albatrosses, get caught as by-catch in long-line fishing operations. As many as 100,000 albatrosses are hooked and drowned in massive long-line fisheries operations every year.

Indeed, birds connected with oceans, with the exception of many sea gulls, are generally in trouble. For example, nesting colonies of seabirds in the North Atlantic are suffering extremely high chick mortality. Not long ago Flatey Island off northwest Iceland was alive with murres, puffins, kittiwakes, fulmars, skuas, razorbills, and more during the breeding season. Roughly half of all Icelandic seabirds bred there in vast mobs; now only a relatively few individuals remain. In the breeding season the island is coated with abandoned nests, sometimes with eggs in them, but not with chicks, as they once were. Breeding colonies elsewhere in the North Atlantic are also dwindling. There has been virtually no breeding in the largest puffin colony since 2005, and arctic tern colonies have suffered massive chick die-offs. There are multiple probable causes—ocean acidification, climate change warming the seas and generating unseasonal storms, and the building toxification of the planet. All these changes affect the oceanic food chains on which the breeders depend and the conditions in nesting colonies. Toxics are especially problematic, as they are carried northward by currents and winds from north-

Albatrosses are threatened by longline fishing, which often hooks them, by depletion of the populations of fishes on which they feed, by plastic and toxic pollution, and by climate change. Grey-headed albatrosses primarily feed on squid and are a species in rapid decline. This individual is shown with a chick on a typical albatross's raised mud nest.

ern hemisphere industrial areas. Winds and currents sweeping pollution northward from Europe, North America, and China bring more bad news for seabirds preying high on the food web. Mercury, which affects avian behavior and breeding, is widespread and increasing rapidly in some places. Pesticides, polychlorinated biphenyls (PCBs), non-stick perfluorinated coatings (PFCs such as Teflon), flame retardants, and plasticizers are often found in seabirds, as are chemical-laced microplastics.

Human activity is clearly pushing some seabird populations toward extinction, not just in the North Atlantic, but also in the tropical Pacific, where the worst impacts have been from the transport of animals, especially the Polynesian rats spread by the first human invaders, and later by black rats that accompanied Europeans. Domestic cats, those major enemies of bird fauna, pigs, and other domestic animals have done their bit too. The most recent estimate is that the Pacific avian holocaust exterminated about 1,300 species but because of a poor fossil record the number could be as high as 2,000. The results of the holocaust are obvious. For instance, both Tahiti and Pitcairn Island are loaded with rats and other exotics. Tahiti has disturbingly bird-free tropical forests, with a handful of struggling endemic birds surviving from a once relatively rich fauna. Pitcairn is similarly infested, with a single surviving terrestrial endemic, a reed warbler, being amazingly common. But seabirds such as shearwaters and petrels, which nest in burrows, have been extirpated from many, if not most, islands occupied by people.

Two islands in the Pitcairn group make an interesting comparison. Henderson Island is a 36-square-kilometer (14-square-mile) "makatea"—a chunk of uplifted, mostly razor-sharp coral. Much of Henderson is bordered by cliffs 12.5 to 15.6 meters (40 to 50 feet) high; access is difficult, and the only freshwater source is a spring below the high-tide line. For these reasons it is no longer inhabited. It has four surviving species of endemic land birds, some endangered, but no significant seabird nesting. It was infested with rats, and in 2011 there was a multimillion-dollar attempt to exterminate them, but sadly it failed. The nearby island of Ducie, in contrast, is a very low uninhabited atoll, free of rats. It is famous as a seabird breeding ground, supporting a giant mob of birds that includes some 95 percent of the world's population of Murphy's petrel, some 5 percent of the Christmas shearwaters, and many others. Ducie is already vulnerable to inundation by storm surges and will soon be disappearing under the waves. The hope is that the birds will be able to move to and breed on Henderson, but that depends on whether the latter can be cleared of rats, an expensive and extremely difficult operation.

The problems for birds associated with oceans are not confined to seabirds. In the United Kingdom wintering wading birds are disappearing from estuaries. Redshanks, ringed plovers, oystercatchers, curlews, and dunlins are declining rapidly—possibly because wintering grounds are shifting northeastward in response to global warming or because of failing success on arctic breeding grounds. The failure of breeding could be related to changes in the timing of bird migrations and insect emergence times, or even to the increasing pollution of the Arctic by long-lasting toxic chemicals mentioned above. In most cases there are inadequate data on the ecosystem effects of those poisons, and none at all about the combined (and likely synergistic) impacts from the diverse poison cocktails to which the birds are exposed. With oceanic systems being altered by climate disruption, ocean acidification, exposure to toxins, plastic garbage, and increased sedimentation, the predicament of oceanic birds may be even more dire than that of their land-dwelling relatives.

No Time to Waste

As our sample of today's threatened and endangered bird species suggests, thousands of species of birds are in retreat. Populations are disappearing on a weekly basis, and every year the list of extinct species grows. Some birds, of course, are doing fine. Invasive species, such as starlings in North America, are thriving. These aggressive passerines easily outcompete native species for nest holes, a population limiting factor for them. Rock doves (pigeons) and English sparrows are abundant in many parts of the world that their ancestors never occupied; so too are many aggressive gull species. People and their garbage provide habitats and resources for the gulls, transforming what were once just one of nature's many seabirds into little more than "flying rats." But these are some of the major exceptions to the general pattern of ever-declining avian populations.

Hope is hard to muster. The past doesn't present a pretty picture on the bird extinction front, and neither does what's happening right now. In both temperate and tropical regions, birds typical of open areas are slowly disappearing. In many places migrant North American songbirds that dwell in forests are heard singing less and less. Birds that call the water home are fading away. The best estimate is that 1,313 (13 percent) of the known 10,027 bird species existing are currently classified by the International Union for Conservation of Nature (IUCN) as threatened with extinction, and another 880 (9 percent) are classified as near-threatened. Of the remaining 78 percent, too many populations are declining, if not as precipitously as the least fortunate species. If the environmental trends of today continue, with little or nothing being done about human population growth, destruction of natural areas, careless transport of species, rising consumption, and profligate releases of greenhouse gases and toxins, many reputable scientists believe one-third of all bird species, and an even greater proportion of bird populations, will be gone by the end of this century. The future of those birds—and innumerable other creatures—is in our hands. Our horrendous ethical and esthetic sin is one that could be assuaged if we as a species dedicated ourselves to preserving—instead of destroying—nature.

The nineteenth century saw humans spreading across the globe and mammals falling in our wake. These bison skulls are a reminder of how, in the absence of a conservation ethic, this once widespread species almost disappeared at the hands of unchecked exploitation.

6 MAMMALS LOST

Sometimes called the goddess of the Yangtze, the baiji was one of very few species of freshwater dolphins. It was endemic to the Yangtze River in China, where it disappeared around 2007. It is the first marine mammal that has become extinct in the past fifty years. Its extinction is a sad reminder of our massive assault on nature and of the sad state of freshwater creatures worldwide.

MANY YEARS AGO, the American naturalist William Beebe wrote: "The beauty and genius of a work of art may be reconceived, though its first material expression be destroyed; a vanished harmony may yet again inspire the composer, but when the last individual of a race of living things breathes no more, another heaven and another earth must pass before such a one can be again."

Indeed, Beebe understated the case; new living beings may eventually evolve that resemble those that have been lost, but they will differ in many ways from the original versions. Among the multitudes of animals that have disappeared forever, along with the ancient trilobites and the giant dinosaurs, is a long list of mammals extirpated by *Homo sapiens*—a grim reminder of the pervasive negative impact of human activities on Earth's biodiversity.

Among life's several kingdoms, one, Animalia, contains several phyla, and in one of them (the Chordata) is the subphylum of vertebrates. Among the vertebrates are several classes, including the great class Mammalia. We are one species of that class, which contains those animals that have hair and suckle their young. In our home class are more than 5,500 species, and they are, taxonomically speaking, our cousins. We mammals have a long and rich history, having evolved more than 200 million years ago from a group of reptiles known, aptly enough, as the mammal-like reptiles.

Given our kinship, there is something especially poignant about the plight of the mammals. Strong historical and scientific records tell us that, just since the year 1500, roughly eighty mammal species have passed through the gates of extinction. An additional twenty-seven are "possibly extinct," meaning that they are very likely extinct but have not yet been given that status by the International Union for Conservation of Nature (IUCN), the world conservation governing body that declares their passing. If you now want to see specimens of these 107 species, you'll have to look in the British Museum of Natural History in London or the Smithsonian Institution's Natural History Museum in Washington, D.C., where you might find woeful reminders of the finality of extermination.

Early extinctions occurred on islands and in temperate regions such as northern Africa, Europe, Central Asia, North America, and South Africa. In the past two centuries most extinctions of mammals have occurred in Australia. At present, there are large concentrations of endangered mammals in tropical regions of the world, including the Andes in northern South America, the Mata Atlântica in eastern Brazil, the Congo basin in West and Central Africa, and Southeast Asia, especially in Vietnam, Cambodia, the Malay Peninsula, and the Indonesian islands.

Australia's Disgrace

Australia is the continent where by far the largest number of mammalian extinctions has occurred. It has also become a textbook example of the ravages caused by introduced animals and habitat destruction. Most native Australian mammals are marsupials, which give birth to undeveloped young that continue their development inside their mother's ventral pouch. Because few predators were native to the continent, many marsupials lack strong defensive mechanisms.

Although many marsupials successfully coexisted for tens of thousands of years with Australian aborigines, the arrival of European settlers in the 1700s changed everything. The deliberate introduction of alien predators such as foxes and cats, the accidental introduction of omnivorous rats, and the arrival of competing herbivorous species such as rabbits and camels coincided with the accidental introduction of pathogens. Added to this set of problems was massive destruction of natural habitats. Appreciation of the fragility of Australia's ecosystems and the vulnerability of its marsupial fauna came too late. The result was the historic extinction of at least twenty-seven species of native Australian mammals.

The extinct mammals include the inconspicuous broad-faced potoroo, a rat-sized marsupial that was first described in 1844. It was already very scarce when the Europeans settled in Australia and became extinct in 1875. The eastern hare wallaby, a small member of a group of marsupials that look like miniature kangaroos, was found in southeastern Australia until it became extinct in 1890. The causes of its extinction are unknown, but it is presumed that habitat destruction and predation by cats caused its demise.

Fires, habitat conversion to agricultural fields, and grazing by domestic animals became more widespread in the twentieth century and caused the extinction of many more marsupials, such as the Toolache wallaby, last seen in 1924, and the lesser bilby, a rabbit-resembling marsupial with very large ears that went extinct

The Tasmanian tiger, with its remarkable coloration and tiger-like stripes, was the largest predator marsupial. Females were unique in that their pouch opened to the rear and, interestingly, the males also had a pouch into which they could withdraw their scrotum. The last captive individual died in 1936.

in 1931. The bilby was found in the deserts of central Australia and was probably eliminated by introduced predators. Some other species were not discovered until after they went extinct, such as the central hare-wallaby, known from a single skull discovered in the Lake Mackay area on the Western Australia–Northern Territory border in 1932. The crescent nail-tail wallaby was a beautiful, aptly named species, with dark, yellowish, and white color bands on the face. It lived in forest and scrubland in central and western Australia. It disappeared in 1956, probably because of predation by red foxes and destruction of its habitat by the spread of agriculture. The desert rat-kangaroo once lived in sandy habitats in northern Australia. It was thought to be gone at the end of the nineteenth century but was rediscovered and lost again several times afterward. Finally, in 1994 all hope was gone and it was declared extinct.

One of the most tragic of Australia's extinction events is that of the thylacine, more often referred to as the Tasmanian tiger. It was the largest carnivorous marsupial surviving into modern times, similar to a coyote in size and shape. On the one hand it was ferocious, but on the other hand nurturing—rearing its young in its marsupial maternal belly pouch. Thylacines became extinct in mainland Australia by the time of European colonization but persisted on the Australian island of Tasmania. The species probably persisted in Tasmania for a while, possibly into the 1960s, but is now almost certainly gone despite occasional "sightings" up to the present.

Thylacines often killed sheep and chickens brought in by settlers, so the Tasmanian government encouraged residents to hunt them and even paid a bounty. In the late nineteenth century, a naturalist wrote: "the native dog, is a marsupial animal . . . covered with a dirty yellowish brown fur, with traverse stripes of a brownish black colour on its back. These animals caused much annoyance to the first settlers . . . [so] it was found necessary to offer a reward for destroying

Steller's sea cow was a huge species, slow, tame, and full of fat. This kelp-feeding species of the "sea cow" order (Sirenia) was discovered in 1741 and hunted to extinction in less than thirty years.

them." The extermination effort proved to be quite efficient and was augmented by various factors, including introduced diseases from domestic dogs and the destruction of native habitat. The last lonely Tasmanian tiger died in 1936 in a zoo in Hobart, the capital city of Tasmania.

A Giant Denizen of the Bering Sea

Steller's sea cow, a close relative of the manatees, was a huge animal. Adults weighed more than 8,000 kilograms (17,600 pounds) and measured up to 8 meters (26 feet) in length. The species was discovered by perhaps the most ambitious scientific expedition of the eighteenth century, led by the explorer Vitus Bering and sponsored by the Russian Academy of Sciences. The expedition's goal was to explore the lands and waters of the Russian Far East.

In 1741 numerous problems beset the expedition, and illness afflicted many of the crew. Bering himself died that year on what would later be known as Bering Island in the Bering Strait. Another twenty-eight members of his crew also perished on that barren piece of land. The doctor-naturalist on board was Georg W. Steller, who discovered the sea cow that bears his name. This stunning animal was apparently very abundant in the cold waters near the island. Steller wrote that there were "quite the many Manatee near shore in the water, which I had before never seen," and also mentioned that "these animals are fond of shallow sandy places along the seashore, but they like especially to live around the mouths of rivers and creeks, for they love fresh running water."

In the 1700s the sea cows were already restricted to a small geographic range. Later research indicated that the populations Steller discovered were a relic of a larger distribution that had existed a few hundred thousand years earlier. But Steller's discovery marked the beginning of the end for the huge sea cows. Subsequent expeditions indiscriminately hunted the virtually helpless animals for their meat, fat, oil, and hides. A

Like the much-larger Steller's sea cow, the Caribbean monk seal was relentlessly hunted to extinction. But unlike the sea cow, it lasted more than four hundred years after its discovery by Christopher Columbus, despite its occupancy of warm waters much closer to large human populations.

naturalist on an expedition in 1802 wrote, "Sea-cows were formerly frequent at the coast of Kamchatka and the Alëutian Islands, but in the year 1768 the last animal of this species was killed, and since then none has been seen any more." That last individual was killed only twenty-seven years after Steller discovered the species.

The Seal of Extinction

Steller's sea cow is one of many marine mammals human beings have wiped out. Another is the Caribbean monk seal, a species whose extermination was begun by Christopher Columbus and his crew. The seal was once widespread, living along the shores of the Caribbean Sea and the Gulf of Mexico, ranging from Texas to the Yucatán Peninsula to Honduras and around islands such as Jamaica and Cuba. It is the only species of seal, so far, to have been wiped out by people.

Columbus discovered this large seal in 1494 during his second voyage to the Americas and immediately ordered it to be hunted. His crew took eight seals and cut them up for oil and meat. That was the beginning of almost five hundred years of relentless persecution. For a long time, particularly in the first two or three centuries, the species was exploited as a food source and for the products that could be made from its remains. Later it was also killed because it was seen as a competitor to fishing operations. The last reliable record for this seal was in 1952, from the Serranilla Bank, a semi-submerged reef in the western Caribbean between Nicaragua and Jamaica. In 1986 an article by Berney LeBouef and colleagues appeared in the journal *Marine Mammal Science* with the stark title, "The Caribbean Monk Seal Is Extinct." LeBouef and his crew had searched the scientific literature for articles, looked through historical records, and visited every reasonable location where the monk seal might be found. Their article extinguished the last faint hopes that a small remnant population might still exist somewhere.

A Freshwater Dolphin

The most recent extinction of an aquatic mammal, and the first dolphin to be eliminated by human activities, was the baiji, or Chinese river dolphin. The baiji was a rare breed, one of only six freshwater dolphins that are known from modern times. The remaining five species are all categorized as endangered. Various species of these unusual aquatic mammals inhabit large rivers in Asia or the Amazon basin.

Living in the murky waters of the Yangtze River, the longest river in China, the baiji was almost blind and navigated by echolocation, as bats do. A beautiful creature, approximately 2 meters (6.5 feet) long, its numbers have declined in the past five decades as a result of the rapid development in China. Pollution, drowning in fishermen's nets, overfishing, damming of rivers, and the diminishing volume of water in the Yangtze are some of the factors causing its demise. By the late 1980s only an estimated four hundred baiji remained—not a lot, but still a large enough number that they could have been saved. But in the 1990s the population dwindled to around one hundred animals.

In 2006 an international team of scientists with sophisticated technological instruments conducted a futile search along more than 3,500 kilometers (2,175 miles) of the Yangtze River. The lead naturalist reported: "When we started, we were really optimistic about finding them, but as each day went by it became increasingly clear that there are no baiji left." He also wrote about watching a film of Qi Qi, a baiji that had lived twenty-two years in captivity until its death in 2002: "I consider myself a strong man, but when I saw that footage [of Qi Qi] I cried for several minutes. It's just so terribly sad." What was almost certainly a lone baiji was filmed in 2007, but it was merely an echo of a species that had numbered about five thousand individuals as recently as the 1950s. The species was declared extinct in 2008.

In modern times, people usually don't give up easily on disappearing species, especially dolphins. The baiji was no exception. This amazing animal became a conservation cause in China, and dramatic efforts were made by Chinese and international scientists, conservationists, and politicians to save it. But the lethal counteracting force of economic growth is a powerful one. Species like the baiji become icons when they are so rare that saving them is almost a miracle. Attempts to pull off this miracle are very expensive, and the economic forces arrayed against it are typically so entrenched that political and business interests deem the tradeoffs unreasonable. For many the choice becomes dolphins or jobs. And so a species vanishes, tears are shed, and "progress" marches on.

Przewalski's horse, a close wild relative of the domestic horse and tarpan, originally ranged from Europe to China. Its characteristic stiff, erect mane can be clearly seen in this picture. Przewalski's horse was once considered extinct in the wild but was reassessed when a single mature individual was seen. Now, thanks to a successful captive breeding program, the horse has been successfully reintroduced into its native Mongolia, where a small population of some fifty individuals exists. Its perilous state was produced by ongoing threats of hunting, habitat loss, and interbreeding with domestic horses.

The World's Wild Horses

The tarpan was a wild species of horse in the steppes of Central Asia from which domestic horses seem to have originated thousands of years ago. As the human population in the steppes grew, the tarpan became increasingly rare. It was hunted for centuries as a food source, and eventually many were captured for domestication. The last tarpan died in captivity in Poland, probably in 1887. Its genetic echo, most scientists think, lives on in the many forms of selectively bred domestic horses.

Another horse species, the quagga, also became extinct in the nineteenth century. Some biologists considered it to be a full-fledged species and others a subspecies of the plains zebra. Quaggas were found exclusively in southern South Africa. They had a very distinctive color pattern, with almost no black stripes on the posterior part of the body; their stripes were only visible on the head and neck. Quaggas were hunted to extinction in the nineteenth century, as were the populations of many other animals in South Africa. The last wild quagga was killed in 1879, and the last individual in captivity died in 1883. There are a few complete specimens on display in some natural history museums—more sad reminders of a species that did not survive long enough to see the birth of the conservation ethic.

Darwin's Fox

When Charles Darwin visited the Malvinas (Falkland) Islands in 1834, he found and collected specimens of a new species of fox, which has become known as the Falkland island fox, or warrah. The foxes were relatively abundant and quite large, about twice the size of the familiar red fox. During the summer, they dined on the seabirds that formed huge nesting colonies on those islands, but how they managed to survive the very harsh winters is a mystery. Warrahs were unbelievably tame. Darwin collected his specimens by simply hitting them with a stick. In the notes of his famous voyage around the world on the SS *Beagle*, he wrote that warrahs were so naïve about people that he was afraid they would soon become as extinct as the dodo. Darwin was right; the last individual was killed around 1876 because it preyed on introduced

The quagga, a denizen of the grasslands of southern Africa, has been extinct since the late nineteenth century. Its name, a bushman rendering of its call, is onomatopoetic. This interesting zebra was apparently exterminated by hunting for meat and hides and just plain bloodlust (it was seen as a competitor of domestic grazers). So we can only appreciate its unusual and unexplained restriction of striping to the forward end of its body by seeing it, as here, on stuffed specimens in museums.

domestic animals such as sheep, becoming an enemy of settlers.

The Rats of Christmas Island

Until recently, the cause of the extinction of the two endemic rat species from Christmas Island, off Australia, known as Maclear's rat and the Christmas Island rat, was thought to be known. The story went like this: in 1899 a ship landed at the island, and a non-native species, the Polynesian rat (also called Pacific rat or kiore), found its way ashore. The new invader then outcompeted the natives, increased in numbers, and the two native species went extinct. It turns out, however, that the story is much more complex.

The invasive Polynesian rat was indeed the cause but, it has been discovered, infection more than competition was the culprit. The introduced Polynesian rats were infected with a microscopic parasite, which is related to the parasite that causes sleeping sickness (nagana) in humans. The Christmas Island rats had no resistance to the new parasite, causing the native populations to decline rapidly. Competition with the Polynesian rat was likely also a problem, but secondary to the infection. The last individuals of either native rat species were seen in 1903.

Flying Mammals

Bats, the only true flying mammals, comprise a group that faces enormous challenges in modern times. But their troubles go back many decades, even centuries, and ten bat species have become extinct since the days of sailing ships. One of the early known bat extinctions involved the Mauritian flying fox, which was native to the islands of Mauritius and Reunion in the Indian Ocean. In 1772 a naturalist provided a detailed description of the habits and destruction of these large bats. This description occurred two decades after the official scientific description of the species was published:

> *When I arrived these animals were as common, even in the settled areas, as they are rare today. They are hunted for their meat, for their fat, for young individuals, throughout all the summer, all the autumn and part of the winter, by whites with a gun, by negros with nets. The species must continue to decline, and in a*

short time. In abandoning populated areas to retreat to those that are yet to be so, and into the interior of the island, fugitive negros do not spare them when they can get them... One never sees them flying by day. They live communally in the large hollows of rotten trees, in numbers sometimes exceeding four hundred. They only leave in the evening as darkness falls and return before dawn... I have seen the time when a bat-tree... was a real find. It used to be easy, as far as one can judge, to prevent these animals leaving, then to take them out alive one by one, or to suffocate them with smoke.

The Mauritian flying fox was last seen in 1873, another victim of hunting and habitat destruction.

Among the other bat species that have disappeared is the Guam flying fox. This small bat occurred on Guam and the Marianas Islands in Micronesia. The locals hunted it because it was considered a delicacy, but it also suffered from the introduction onto the islands of the brown tree snake. Discovered scientifically in 1931, the species often roosted with another species that is larger and much more abundant, the Marianas flying fox. Forty-six years after its scientific discovery, the last known specimen, a female, was found roosting at Tarague cliff in March 1967. None have been spotted since then, despite extensive searching. As a result, the species is considered extinct.

The most recent bat extinction that has been recorded occurred on Christmas Island, off Australia. A small insectivorous bat known as the Christmas Island pipistrelle became officially extinct on September 8, 2009. The Australian government announced on that day that attempts to capture the last known individuals for a captive breeding program had failed.

Deer and Gazelles

In the past few hundred years, many populations and several species of deer have become extinct, in large measure due to hunting. One was the Schomburgk's deer, which was described as a beautiful animal. It was apparently endemic to central Thailand, although some inconclusive evidence suggests that the species may have been more widely distributed and existed also in Laos. We know very little about how it lived, but it seems to have inhabited seasonally flooded swampy plains, as do other deer such as the barashinga or swamp deer in India and Nepal and the pampas deer in South America. The last confirmed wild Schomburgk's deer disappeared around 1932, although a few captive individuals survived until 1938.

The Saudi gazelle was once found on the Arabian Peninsula. One of the few large mammals found in that region, it was hunted to extinction. It lived in small groups of up to twenty animals on the sandy arid plains from Kuwait to Yemen. It was a pretty animal that weighed up to 30 kilograms (66 pounds). There is no solid evidence related to when it disappeared, but it was declared extinct in the wild in 1980 and officially extinct in 2008. The sandy plains in the Arabian Peninsula are now emptier than ever.

The West African Black Rhino

Of the many populations of mammals that have become extinct in historic times, such as the Hokkaido wolf, the Caspian tiger, the Merriam elk, and the

The black rhino is familiar to almost everyone. Despite attempts at protection, rhinoceroses may be the most endangered big mammals on the planet. They are most threatened by poaching because some cultures think rhino horns have magical properties—especially as an aphrodisiac or a cancer cure. It was thought that the development of Viagra would help save the rhinos. Recently, however, it has been rumored that criminals have been mixing Viagra with powdered rhino horn to help maintain prices that can approach $50,000 per pound.

Syrian wild ass, for us some of the saddest news we have heard came in the form of a press release in April 2013. Issued by the International Union for Conservation of Nature (IUCN), it announced that the West African black rhino was officially extinct. The distribution of this subspecies of rhino had been reduced in recent years to a small area in Cameroon. The declaration was made after unsuccessful efforts to locate any survivors and confirmed the bitter news of the fate of this remarkable animal. This subspecies had survived precariously on the brink of extinction for many years, but finally succumbed to poachers. It was one of many victims of the well-organized international criminal groups that slaughter rhinos to sell their horns on the black market in Southeast Asia and China. The rhino horns are used for fruitless concoctions based on unfounded beliefs about aphrodisiac powers to be gained through their consumption.

Saving the rhino and all other endangered mammals is a gigantic task that has to be undertaken, regardless of its costs and efforts. The fate of ourselves is at stake just as theirs is. The future will be implacable in its judgment of our actions.

The tiger is perhaps the most awesome predator on Earth, although it devours many fewer of the most destructive animal, *Homo sapiens*, than do crocodiles. It may soon be extinct in the wild, in which case it will need to depend on that prey animal to preserve it in zoos. Fortunately, tigers are sufficiently uninhibited to reproduce happily in captivity, but their fecundity ironically creates a problem for zoos. Tigers require a lot of space, and adults need to be separated from their fast-growing offspring when the latter are about 2 years old, so zoo populations must be carefully controlled.

7 VANISHING MAMMALS

Vaquitas become entangled in illegal gill nets used to capture totoaba fish, whose gallbladders bring thousands of dollars on the Chinese black market. Fewer than one hundred survive, and if the killing continues, they will become extinct in less than ten years.

For some mammals the bell has not yet tolled, though it may before long. The most recent International Union for Conservation of Nature (IUCN) list of animal species verging on extinction indicates that at least 188 mammals are critically endangered, 450 are endangered, and 493 are threatened with extinction. In other words, one in five of all wild mammal species are considered at risk of extinction. Indeed, many more mammals may be at risk, but we lack enough information about their status to know for sure. This chapter tells the stories of a few of those mammals unlucky enough to be in the 20 percent that are threatened or endangered.

The Little Porpoise

The Upper Gulf of California of Mexico is a remote area that hides a marine treasure, the vaquita, a very small porpoise found exclusively in the northernmost 5 percent of the Gulf, with the smallest range of any marine mammal. It is the most endangered marine mammal in the world. Compared to more typical dolphins and porpoises, the vaquita is an oddity. At 1.5 meters (5 feet) from nose to tip of tail (or more properly, from rostrum to fluke) and weighing just 50 kilograms (110 pounds), it is considerably smaller than its fellow cetaceans (whales, dolphins, and porpoises). A hunter of fish and squid, it spends most of its time in shallow waters.

The vaquita has one major problem—getting entangled in fishing nets. The little porpoises are caught by accident and drowned, recently at a rate of thirty or more a year. For some species thirty deaths a year would not put a dent in the population, but vaquitas are down to the last eighty individuals in the wild, and none exist in captivity. In a race to save them, many actions have been taken to protect them. Notably, the region they inhabit has been declared a protected area by the Mexican government, and restrictions on fishing in the vaquita's habitat have become law. Also, fishermen have been paid to change the type of nets they use in areas where the vaquitas are found. Illegal use of the old nets continues, however, because not enough effort has been made to help local communities and fishermen move from destructive entangling nets to other fish-capturing methods. If that problem can be solved, the conservation efforts may save the species, if it can hold on long enough. As recently as May 2013 the Mexican government announced a very strict conservation program to try to save the vaquita. Full adoption of the new kind of net that does not entangle the vaquita is the last hope for its survival.

Giants of the Seas

Winter days in the waters off the Baja California peninsula of Mexico are usually cold, with clear skies. Winds sometimes howl day and night. The shallow waters of the coastal lagoons in the Pacific Ocean host a massive influx of gray whales in that season, arriving to give birth after a long migration from the cold Arctic waters.

Gray whales are abundant today, with an estimated population of 15,000 animals. But they represent a conservation success story because they were close to extinction at the beginning of the twentieth century. In the 1800s American, Russian, and Japanese whaling ships slaughtered thousands of gray whales. The famous whale hunter and naturalist Charles M. Scammon, who was one of the first western explorers of the Baja California estuarine lagoons, hunted more than two hundred whales in the winter of 1858 in the Ojo de Liebre lagoon. That and other lagoons in Baja California were decreed as whale sanctuaries by the Mexican government at the beginning of the twentieth century—a decision that likely saved these giants from extinction.

Many other species of whale were hunted down almost to extinction for oil, bone, and other products in the nineteenth and early twentieth centuries. Baleen whales were particularly subject to overexploitation and, although protected for the past several decades, most species still have reduced populations. The numbers of whales slaughtered in little more than a century are staggering. More than 2 million whales are estimated to have been killed in Antarctic waters alone. Huge whaling camps were established on islands such as South Georgia, the Maldives, and Tierra del Fuego in Chile.

In 1986 the International Whaling Commission imposed a moratorium on commercial whaling, which has significantly reduced the pressure on many whale species. Nevertheless, some countries, particularly Norway, Japan, and Iceland, still hunt some whales every year for what they claim is scientific research. In March 2014 the International Court of Justice declared Japan's whaling illegal, giving hope that whales may roam freely in the oceans again.

Gray whales became almost extinct in the northern Pacific in the late 1800s because of hunting. In 1917 a Mexican president decreed the protection of this and other marine mammals in Mexican waters. Slowly the populations recovered, and the whales are now relatively common.

Even after almost three decades since the commercial whale-hunting moratorium was established, several species are still at risk because of the past impact of human activities. Perhaps the most endangered is the northern right whale, found in the waters of the North Atlantic, where Russian and Norwegian vessels systematically hunted it; fewer than three hundred individuals survive. Similarly, bowhead whales in the Arctic are also still endangered by past and current effects of human activities, with a population of fewer than one thousand. Blue whales, the largest animals (nearly 30 meters, or 100 feet, long) ever to roam this planet, have probably been reduced to fewer than 3,000 individuals. The status of smaller whales, such as the Peruvian pygmy beaked whale, is less well known, but apparently they are very rare.

Aside from hunting, whales are being directly or indirectly impacted by other human activities. We saw that little Laysan albatross chicks are being killed by being fed some of the gigantic amounts of plastic debris now floating in the oceans. But at the other end of the size scale, whales too are now dying from ingesting plastic trash. This is especially a problem for sperm and beaked whales since plastic bags, sheeting, and other intestine-blocking objects often resemble natural prey items such as squid. But baleen whales are hardly immune, even though they eat plankton. They take in huge amounts of water when they feed, swallowing with it the refuse of our society. In 2010 a gray whale died after stranding itself on a Washington state beach. In its stomach it had a golf ball, surgical gloves, duct tape, miscellaneous plastic fragments, a pair of sweat pants, and twenty plastic bags, among other trash. Plastic cannot be digested and simply clogs the gut, causing death not directly but indirectly, through starvation and disease.

Increased industrial noise in the oceans, from the engines of ships to air guns used in seismic exploration for gas and oil to military sonar tracking and bombing practice, is now known to cause many problems for marine creatures. Primarily, it can be very dangerous for whales and other cetaceans, many of which use sound for communication and may use blasts of sound in their hunting. Noise has been implicated in many cetacean strandings and deaths, although precise data are difficult to collect. Of course, besides whaling, whales, dolphins, and porpoises are also endangered by accumulation of persistent organic pollutants, ship strikes, accidental entanglement in fishing gear, and, of course, like almost all animals including us, by climate disruption.

The Last Tigers

Ranthambore, a bustling city of more than 5 million people, is southwest of New Delhi, India's capital city. It is also the guardian of one of the natural gems of India and the whole planet: the Ranthambore National Park. In the park lives a population of wild tigers, one of the last remnant groups of what once was one of the most widespread big cats in the world. The park is sort of an island, one of a couple dozen tiger reserves scattered throughout the species' former range. Besides a handful in India, others are in Russia, Bangladesh, Nepal, Bhutan, Laos, Cambodia, Vietnam, Malaysia, and Indonesia. At the beginning of the twentieth century, various subspecies of tigers were still found throughout a vast territory of millions of square kilometers across Asia. The range extended from Mesopotamia and the shores of the Caspian Sea all the way to China, Korea, and the Kamchatka Peninsula in eastern Russia, and from India, Nepal, and Bhutan to the Malay Peninsula and the islands of Singapore, Bali, Sumatra, and Java.

As human populations swelled in Asia throughout the twentieth century, the pace of clearing forests and killing tigers increased. Over a single century, the great beasts became increasingly rare, as trophy hunting, poaching for the mythical medicinal properties of their body parts, habitat destruction, and predator-control measures combined to make life outside tiger reserves virtually impossible.

The species' power is revealed by the largest of its members, Siberian tigers, which weigh as much as 320 kilograms (710 pounds). Bali, now tigerless, once had the most diminutive tigers, but they still weighed up to 100 kilograms (220 pounds). Wherever they roamed, tigers were formidable predators with little fear of humans. Indeed, some have become habitual man-eaters. Jim Corbett, a British hunter and naturalist, became famous for killing tigers and leopards that had added people to their regular diets. He killed the Champawat tiger, a female that was reputed to be responsible for 436 human deaths. That tigers in the region killed 436 people is documented, and some reasonable evidence supports that this tiger was solely responsible.

On the island of Singapore, only 624 square kilometers (241 square miles) in extent, tigers reportedly killed many people every year before they were exterminated in the 1930s. Even today, numerous unlucky individuals are killed or mauled by tigers in India. One hotspot of tiger attacks is the Sundarban mangrove forest in the vast delta on the Bay of Bengal, where the Meghna, Brahmaputra, and Padma Rivers meet between India and Bangladesh. Every year tigers kill dozens of people in this locale. Being a handsome, powerful, and feared animal makes its conservation a tricky issue. From the safety of distance, they are admired, but in their neighborhoods human sentiments are more complicated.

Classifying the different types of tigers has always been difficult. Traditionally, differences in color, size, and location were used to distinguish nine subspecies. More recent genetic analyses indicate that there may be eighteen definable groups, but the groupings do not correspond exactly to the ones described by traditional means. One recently named tiger subspecies from the Malay Peninsula was described based on genetic information in 2004, with arguments that its uniqueness should afford it more conservation attention. In recent times, at least three tiger subspecies have been lost, and another one is probably also extinct. The Bali tiger disappeared in 1937, followed by the Caspian tiger in the 1950s and the Javan tiger in 1976. The South China tiger has not been sighted in more than twenty-five years.

At Rathambore National Park, the clash between modern civilization and nature could not be more dramatic. The boundary between the protected area and urban sprawl is especially stark near the Old City park entrance. One can feel lost when driving through the chaotic winding streets of the Old City, a decaying ancient urban settlement teeming with countless people, domestic animals, cars, carts, and bicycles. The city is bisected by a smelly river that looks more like a filthy sewer, where dogs and pigs fight for scraps of rotting food. The feeling of strangeness is powerful as mobs of people, vehicles, and animals compete to pass through the narrow streets. With that scene still in view, one comes to a rock wall to the left of the main street. Beyond a narrow entrance, lacking any identifying signs, is a completely different scene. Low hills lie directly in front of you, denuded of vegetation, and some people can be seen gathering the last plant remnants as fuel for their fires.

The miserable dirt road continues for a bit until you reach a magnificent centuries-old gate. At that point everything changes. Beyond the gate are rolling hills covered with a mix of grassland and mature tropical dry forest. Wildlife is everywhere. One can see massive nilgai, the largest Asian antelope, and small, attractive (and endangered) Indian gazelles, or chinkara. Quickly coming into view are southern plains gray langurs, one of the most beautiful and widespread monkeys in India. A few hours spent here will reveal literally hundreds of species, including very large herds of chital deer and a few sambar deer. With some luck, the elusive tiger may be seen or heard. A 1997 census in the park estimated that around 4,500 chital deer, 3,000 sambar deer, and 2,000 nilgai lived in the park. Such plentiful prey is the key to having a healthy tiger population. Ranthambore covers a huge area of nearly 400 square kilometers (150 square miles), but even this large park can sustain only a few dozen tigers. Oddly, hunting is what saved the park; it was the hunting ground of the Maharajas of Jaipur. The tiger population hit a record high in 1989 of some forty animals, possibly the maximum carrying capacity of the park. More recently, a population of only about twenty has been reported.

The decline in tiger numbers throughout Asia seems to have a sinister cause: poaching. In the past few years, the combined negative influences of corruption, poorly trained guards, obsolete guns owned by park staff, bureaucracy, and higher prices for tiger parts and bones in China have led to a tiger holocaust. Worldwide, almost one thousand tigers were recorded as killed by poachers from 1994 to 2010, but that is only a fraction of the real number. Tigers are now in critical condition everywhere, getting closer to extinction in the wild day by day. The World Wildlife Fund for Nature (WWF) recently estimated that the global population of wild tigers is about 3,200 individuals, a mere 5 percent of the 100,000 tigers estimated to have been living in the early twentieth century.

Can humanity be forgiven for our apathy? Some will be spared the disdain of future generations, such as the Russian conservationists that the nature writer Peter Matthiessen described in his extraordinary book *Tigers in the Snow*: "In arguing for the heroic efforts on behalf of the tigers, one could cite the critical importance of biodiversity as well as the interdependence of all life . . . But finally those abstractions seem less vital than our instinct that the aura of a creature as splendid as any in our earth, infusing man's life with myth and power and beauty, could be struck from our experience of creation at a dreadful cost."

The global situation for tigers is so precarious that a number of biologists think there will be no tigers left in the wild in two decades. Such a slaughter need not happen, but well might.

The Mighty Lions

In human lore, lions and tigers seem to go together like salt and pepper. Both are big and dangerous. Lions are virtually synonymous with the wild strength of nature. Long ago, when they freely roamed not only in Africa but as far as southern Greece and Italy in Europe to Mesopotamia and India in Asia, they were both feared and respected. Although their huge historic range in Asia has shrunk to a single region in India, until recently lions were still plentiful in Africa. But recent evidence suggests that they are now in jeopardy even there, and their fate may be converging with that of the tigers.

Around 1950 there were an estimated 1 million lions in Africa, with a geographic range extending from the shores of the Mediterranean Sea to the Cape of Good Hope in the southernmost part of the continent. Today there may be as few as 23,000 lions in all of Africa. The West African lion is at the brink of extinction, with only 250 individuals surviving in a few countries such as Nigeria. The species has suffered from habitat encroachment for urban settlements and croplands, poaching, decline in prey populations, diseases transmitted from domestic animals, competition with people for habitats and prey, and direct hunting (because lions kill cattle and sometimes people). One subspecies, the Cape lion, was famous for the dark manes of the males. Cape lions were abundant in the early nineteenth century but were exterminated by 1861. The Barbary lion, originally found from Morocco to Egypt in northern Africa, became extinct by the 1950s.

Lions, like many other species, face new dangers as the human population continues to expand. One story reveals how unexpected problems can devastate a lion population. In a 2010 interview, a guide from Maasai Mara National Park reported that in early 1994 tourists flying in a hot air balloon over Serengeti National Park in Tanzania saw a male lion acting strangely. He convulsed and then collapsed, dying in a few hours. That lion was one of the first victims of a mysterious disease that killed about one-third of the lion population within a year. Other individuals suffered permanent nerve damage. It was later discovered that the lions were afflicted with canine distemper spread by the 30,000 domestic dogs living near the park. Diseases transmitted by domestic cats, dogs, and cattle increasingly threaten wild felines around the world.

The few wild lions outside Africa all live in the Gir Forest National Park and Wildlife Sanctuary in the state of Gujarat in western India. These are members of a subspecies called the Asiatic lion, which is slightly smaller than its African counterpart. Once part of the larger distribution of lions, by the mid-twentieth century the Asiatic lions were almost completely gone. The tiny remnant population in the Gir Forest owes its existence to a viceroy of the province, who afforded them a measure of protection at the beginning of the twentieth century. At the time only fifteen Asiatic lions remained. Since then the Gir Forest lion population has been a conservation success story. Although the subspecies is still critically endangered, approximately 400 lions now live in the 1,280 square-kilometer (482-square-mile) park.

The future of these lions is uncertain, due to the huge human population on their doorstep. The Gir Forest is surrounded by 400,000 people, causing

very complex problems that continually compromise the long-term survival of the big cat. Even extreme measures—such as removing the villages around the reserve and compensating locals if the lions kill one of their domestic animals—may not be enough to prevent their eventual demise as the human population continues to grow. To lessen the danger of someday losing this entire subspecies, plans have been made to reintroduce some Gir Forest lions into the Palpur-Kuno Wildlife Sanctuary in central India.

Although the Asiatic lions of Gir Forest live a mostly wild existence, they are so intensively managed that their behavior resembles that of zoo animals. Most if not all of the lions have been captured at one time or another and treated for wounds or illness, and in times of scarcity their food supply is supplemented. Ironically, they continue to exist in this semi-wild state because of the ministrations of humans—the species that also poses the greatest threat to their future existence.

It is hard to believe that the African lion, the iconic carnivore, may be facing extinction in the wild. Thanks to habitat loss, diseases, and illegal killing, only roughly 23,000 lions remain in Africa's vast savannas—perhaps less than 10 percent of the numbers that roamed there in 1950.

Cheetahs are the fastest land animals, but some of their populations have already vanished. It seems increasingly likely that they will not be able to outrun the human takeover of Earth's natural environments. One hopes that this juvenile cheetah will help maintain the population on the vast African plains.

The Fastest Mammal

Lions and tigers are big and powerful, but speed is the specialty of their relatives the cheetahs. Cheetahs can reach speeds of 100 km/h (60 mph) in a few seconds. They once sped across the dry lands of Asia and Africa, coexisting with tigers and lions when they too were plentiful. Today cheetahs are another of Earth's decimated species and without doubt the most endangered big cats in Asia. The distribution of one subspecies, the Asiatic cheetah, once extended from India to the Arabian Peninsula and Syria and throughout Pakistan, Afghanistan, and Iran. But the conversion of natural habitats into croplands, overgrazing by livestock, depletion of prey species, and heavy hunting overwhelmed the sleek cats. The last Indian cheetah was killed in 1947, and today all that remain of wild Asiatic cheetahs are found in the arid lands of central and northern Iran, where sixty to one hundred are struggling to survive, scattered in several national parks and adjacent areas.

Researchers have determined that the cheetahs of Iran are genetically different from their African cousins. They differ as well in habitat use—Asiatic cheetahs are found in mountainous areas, a habitat very rarely used by their African counterparts. Survival is precarious because the cheetahs share their habitat with livestock and herders. Populations of the natural prey for Iran's cheetahs, such as the Persian gazelle, have been severely depleted. In many places, cheetahs kill sheep and goats, making them a pest in the eyes of herders. Hope for the Asiatic cheetahs lies in the constant conservation efforts of Iranian and international environmentalists.

Compared to Asia, Africa seems like a stronghold for cheetahs. An estimated 12,000 individuals remain, but the species occurs spottily throughout the continent and has disappeared from large parts of its historic geographic range. One reason for the disappearance of cheetahs is the troubling reduction of what has been described as nature's greatest spectacle. In Kenya and Tanzania, millions of wildebeest migrate twice a year, traveling with thousands of plains zebras, topis, hartebeest, and Grant's and Thomson gazelles. They cross the Mara River in an incredibly impressive movement from the southern plains of the Serengeti National Park to the Mara grasslands and then back to the Serengeti. The thundering, sometimes sauntering, herd eats the exuberant grasses that grow after the seasonal rains; when the food is gone, they move on.

This moving ecosystem is made up of plants, prey, and predators. Among the predators are lions, leopards, hyenas, and cheetahs. Whether or not this amazing spectacle will continue is hard to tell. Some have argued that the sun is setting on this extraordinary sight, and thus on the cheetahs and other African predators. Our advice would be to see it while you can, for while hope remains that it will continue, the history of our well-intentioned, if at times inept, conservation efforts suggests the chances are slim.

Sloth Bears

Besides the tiger, one of the great sightings one hopes for in India's Kahna National Park is the sloth bear. At first glance this remarkable animal looks a bit like a big shaggy dog, but a look at the long claws on the front feet will dissuade you of that notion. A second glance might remind you of North American black bears, but sloth bears have a slimmer build, and usually one sees a white V across their chests and a whitish muzzle. The sloth bear is a specialist that feeds mostly on bees and termites. It has a lower lip that has evolved into a device for slurping these tidbits up. But sloth bears will also eat fruit whenever the opportunity arises.

Despite their feeding habits and cuddly appearance, sloth bears can be aggressive and dangerous animals. Park visitors are wise to treat them with caution. In spite of their occasional ferocity, the bears are sometimes used as street entertainment, and reports of abuse are common. Their plight is not a happy one; beyond the disrespectful street performances, their

Giant pandas make great symbols for conservation, but the vast majority of people, including conservation biologists, know them only from photographs and individuals in zoos. Seeing them in the wild like this is a rare treat indeed.

bile is sought by Chinese customers who think it has medicinal value. Finally, their native habitats have dwindled to the point that sloth bears are now listed as threatened with extinction.

Mighty Panda

As a symbol of the need for conservation, no animal has proven as successful as the giant panda—famous as the logo of the World Wide Fund for Nature (WWF). This bamboo-eating relative of bears is highly adapted to its specialized, nutrient-poor diet. It must keep eating large amounts of bamboo, which microorganisms in its gut help it to digest. It saves energy by being sluggish and tubby (so that it has a small heat-releasing body surface relative to its volume). The few thousand pandas still in the wild now occupy a remnant of their former range in southern and western China and northern Myanmar and Vietnam.

Pandas are subject to the usual array of threats: poaching for their soft fur (and for food during twentieth-century famines) and especially habitat destruction. As China's population exploded in the mid-twentieth century, panda survival in nature became problematic despite heroic efforts by the Chinese to conserve them. There are now reserves and an excellent captive breeding program at the Chengdu Research Base, which features artificial insemination since the animals seem to lose their sex drive in captivity.

All this is hopeful, but it is not clear whether that program or the protected areas will prove adequate in the long run. The top panda conservation biologist, our friend Jianguo "Jack" Liu, has been working to improve habitat quality in the important Wolong Reserve, where tourism itself was destroying settings the pandas required. It is worrisome, however, that Jack has never seen a panda in the wild.

The Fading Howl

Little conjures up a sense of raw wilderness like the sound of wolves howling. While sleeping in a nature reserve, imagine being stirred awake by that eerie call. If you are a dog lover, take heart, for the ancestor of your pet is nearby. The gray wolf, according to most research, was the progenitor of our beloved domestic dog. Intimidating as the wolf may seem, it is one of the most misunderstood (and even hated) denizens of our planet.

Excluding domestic species and uninvited guests that live in human settlements, wolves are the most widespread mammal. They have been found throughout Northern Hemisphere landmasses except in severe deserts, ice-covered regions, and tropical forests. They once numbered in the millions, but human persecution has reduced their numbers worldwide to one-tenth of what they were.

Wolves are pack hunters that sometimes prey on livestock, thus naturally gaining the enmity of herding people. Their highly social behavior makes them especially vulnerable to extermination (people can track and destroy a pack of wolves), in contrast to their close relative, the more solitary coyote, whose range is spreading across North America.

In some places, notably the United States and Europe, wolf populations are making something of a comeback from their previous lows. A few decades back, wolves were so rare in the Lower 48 that individuals from populations in Canada and Alaska were reintroduced into the northern Rocky Mountains at the end of the twentieth century. Protected at the time under the U.S. Endangered Species Act (1973), they could not be killed by ranchers, so they quickly multiplied and formed new packs, expanding to as many as five thousand individuals in the Lower 48 by 2014.

Unfortunately, they were perhaps too successful, because they have been delisted as endangered in some states. Although the populations are still monitored and protected to some extent, hunting of wolves has resumed in some of these states, and at least one pack has been wiped out. On the positive side, a lone (and doubtless lonely) radio-tagged young male recently wandered across the Oregon border to become the first wild wolf in California in nearly a century. Not unexpectedly, ranchers did not welcome him.

Gray wolves were common in Mexico until the 1950s. At that time an extermination campaign was launched, sponsored by the U.S. government, which argued that wolves from Mexico entered the United States and killed cattle. Using the powerful 1080 pesticide, thousands of poisoned farm animal carcasses were dispersed in northern Mexico, killing thousands of carnivores from grizzly bears to skunks. Nobody really knows how many wolves were killed, but that campaign started a big decline of the species in Mexico. Hunting and poisoning by hunters continued for the next two decades. By 1970 very few wolves persisted in Mexico.

Ironically, in 1976 the Mexican wolf was declared endangered under the U.S. Endangered Species Act. The United States and Mexico established a binational captive breeding program using the last wild wolves trapped in Mexico between 1977 and 1980. After spending millions of dollars to exterminate wolves in the wild, the U.S. government initiated a very expensive program to save the species from extinction.

The captive breeding program has so far been something of a success, though it has not yet reached its target number of established wolf populations. In 1998 the Mexican wolf was reintroduced to the Blue Mountains in Arizona, many decades after it had disappeared from there. It took more than ten years to reintroduce them in Mexico, but in the late summer of 2011 five Mexican wolves were finally reintroduced in the species' ancient home in the San Luis Mountains in the state of Sonora, near the Arizona-New Mexico border. Unfortunately, by the end of 2012 only one released female was still alive. Nonetheless, the Mexican and U.S. governments are still planning to re-establish the wolf in Mexico. In the spring of 2014 a newly released pair of Mexican wolves produced the first litter born in the wild in Mexico in four decades.

These actions to save the wolves bring hope for what was once a critically endangered species. We will probably never see wolves in the numbers they once had, but at least with this species there are encouraging signs. Humans and wolves, however, are likely always to need conflict resolution. As such, one effort that seems a win-win is compensating ranchers for the cattle they lose to wolf predation. Such an approach lowers tensions, recruits more ranchers into the conservation effort, and is a lot less expensive than driving a species to the brink of extinction and then spending millions more to help it recover. Better ranching practices also protect many species of plants and animals, including some endangered ones. The huge expanse of ranchlands globally presents an enormous conservation potential.

Our Giant Relatives

Gorillas are members of a group to which we belong: the great apes. These fascinating relatives, which are represented by two species (each of which contains two subspecies), live only in the tropical and subtropical forests of Central Africa. For more than a century, gorillas fired the imaginations of explorers, who considered them fierce and dangerous, even though they are usually harmless. True, the threat displays mounted by dominant males are very intimidating, and probably the explorers did not hang around to find out whether the 200-kilogram (440-pound) male's display would be backed up by ferocious action. But among the true dangers of Africa, gorillas are far down the list (mosquitoes actually top the list).

As recently as the mid-1800s, gorillas were known in the Western world only through rumors. Then, in April 1847 Reverend Thomas Savage, a missionary posted in West Africa, was presented with the skull of an unknown ape. He related in the *Boston Journal of Natural History* that "soon after my arrival [to Gabon] Mr. Wilson showed me a skull, represented by the natives to be that of a monkey-like animal, remarkable for its size, ferocity, and habits." Savage immediately saw that the skull was of "a new primate species, one larger and more powerful than humans."

Gorillas are the largest of the great apes and spend their time on the ground looking for food—mostly leaves, stems, shoots, roots, and fruit. In recent decades only one species was officially recognized, but genetic studies have indicated that there are probably two species. One of these, the eastern gorilla, contains two subspecies called the mountain gorilla and the eastern lowland gorilla.

The second species is found in West Africa. The more common subspecies is the western lowland gorilla (which is the gorilla one sees in most zoos), and the other, more elusive subspecies is the Cross River gorilla. Each of the four gorilla subspecies faces enormous challenges resulting from a mix of civil wars, armed roving militias, hunting, poaching for "medicinal" purposes, human-mediated disease transmission (such as Ebola hemorrhagic fever, which is caused by a virus), and habitat destruction.

Mountain gorillas became international news a few decades ago when Dian Fossey was murdered. Fossey was a scientist who had studied the gorillas for more than twenty years and published a widely read book entitled *Gorillas in the Mist*. Her massively publicized murder rallied the international community. The conservation plight of the mountain gorillas became a top priority for many organizations, including the Dian Fossey Gorilla Fund.

Today, mountain gorillas are still rare; estimates of their numbers seesaw, but have never reached beyond several hundred individuals. To see these gorillas is the major reason tourists visit Rwanda, spending quite a bit of money to be guided along the trails of Volcanoes National Park. But the cost and effort are well worth it. Scenes we have observed include a female nursing her baby while a rowdy juvenile frolicked at her feet, and a giant silverback charging to scatter youngsters that had been harassing a baby as it tried to gnaw on the bark of a tree.

The gorilla population suffered during the horrendous genocide in Rwanda that claimed some 800,000 human lives in 1994. The Rwanda National Park guard system broke down during the slaughter, and all Western scientists and conservationists left Rwanda in a hurry. For more than a year, there was no news about the gorillas, and the worst was feared—a subspecies extinction in addition to the human carnage.

The story of the gorillas was both tragic and uplifting. Several local park guards were killed during the civil strife, but a few local conservation heroes managed to protect the gorillas. They received no pay or any other help, but they simply stayed on the Virunga volcanoes to protect their beloved gorillas. When it was finally safe enough for foreigners to return to the park, they were greeted by the ragged, malnourished protectors with big smiles on their faces. Amazingly, when the rebel forces had taken control of the park and surrounding areas, these guards convinced them

of the importance of conserving the gorillas and the forests.

These committed Rwandans were crucial for the mountain gorilla's survival in a very dangerous time, and humanity owes them a great debt. Indeed, more than three hundred people have been killed in the past several decades trying to prevent the poaching of mountain gorillas. The five hundred or so mountain gorillas that live in the small area encompassing parts of Rwanda, the Congo, and Uganda are alive thanks to their sacrifices as well as the bravery of those guards still living. Every day they wake up with one mission: to protect these great apes.

Today the massive volcano range that is home to the mountain gorillas is still covered by forest, but it is completely surrounded by farms. The park is enclosed by a rock fence that protects it against the ever-pressing encroachment of habitat for cropland. From the air it resembles a zoo. The border between the park and the croplands could not be more abrupt. A 90-minute trek from a farm might well bring one face to face with a massive 250-kilogram (550-pound) silverback leaning against a tree. You can look around, see the gorilla, and also view the farm fields and villages below. You can hear barking dogs and faint radio music, a strange mixture of civilization and the last remaining mountain gorillas. It is wilderness of a sort, but one populated by armed guards, guides, and trackers who all protect the gorillas. It is also a tourism success story, providing substantial revenue to both Rwanda and Uganda, permitting some hope that the future might still have a place for these magnificent creatures.

Lowland gorillas are critically endangered by hunting, Ebola virus, and, of course, habitat destruction by an exploding human population. Logging is a big part of this, causing deforestation and creating infrastructure that allows hunters access to remote habitats and a ready market for gorilla meat in logging camps. This male in Gabon may represent one of our close relatives that will be gone by the mid-twenty-first century.

To the west, the situation for the Cross River gorillas is less secure than that of the mountain gorillas. Fewer than three hundred individuals remain, all living in Nigeria and Cameroon. This subspecies is the most critically endangered of the four gorilla subspecies, threatened by poaching and destruction of its forest habitat. At one point the Cross River gorilla was believed to be extinct in Nigeria, but a recent survey indicated that 75 to 110 animals still survive in that country, and an unknown number still exist in the tropical lowland forests of Cameroon.

The western lowland gorilla is the most widespread and abundant gorilla subspecies. In 2008 the Wildlife Conservation Society made a very important discovery when it found a large, previously unknown population living in swamp forests in the northern part of the Democratic Republic of the Congo. With that addition, the current estimated number of western lowland gorillas is from 150,000 to 200,000 individuals, though the populations seem to be declining due to logging and disease.

The most serious disease threatening lowland gorillas is Ebola. This virus is known to have killed many thousands of gorillas around the turn of the millennium. In a remote area on the Congo-Gabon border alone, in the vicinity of the Lossi sanctuary, some five thousand gorillas are estimated to have died. The gorilla epidemic is thought to have been related to a human outbreak of Ebola in the area in the mid-1990s. Although much mystery still surrounds the chief wild hosts of Ebola virus, a fruit bat is now the most likely candidate. It is known that humans have frequently contracted the deadly disease when they butchered or otherwise came in contact with dead primates. It is also evident that hunters, game guides, and tourists can transmit the disease to apes.

As everyone is now aware, Ebola is not just a threat to gorillas, but also to us. The current epidemic is a sign of the deterioration of the human epidemiological environment, tied to factors such as human overpopulation and rapid transportation. At this writing (December 2014) the situation remains serious and the future somewhat in doubt.

Ebola is one added dimension of the problems gorillas face as civilization spreads and wild Africa shrinks. It may sound comforting to hear that there are 150,000 to 200,000 western lowland gorillas in the wilds of Africa, until we learn that our best estimates indicate that the number of gorillas is far less than it was a short three generations ago and may have fallen as much as 80 percent in that time.

Jane's Cause

Chimpanzees don't talk much, don't build airplanes and cars, don't get married, and don't have bar mitzvahs and first communions and all the other things that make our cultures what they are. Yet on the tree of life, they—and their cousins the bonobos (or pygmy chimpanzees)—are our closest living relatives. We are each twigs on the same small branch. We know chimps from movies and circuses, but they are best thought of as the wild animals we appreciate in large part because of Jane Goodall.

At the chimp study site at the Gombe Stream Re-

serve in Tanzania in the early 1960s, Goodall gradually accustomed members of the Kasakela chimpanzee community to her presence. Over time they allowed her to observe their interactions. There were gentle, touching moments, but also savage brutality on display. One day in the early 1970s a female chimp from a different chimp community was brutally attacked by Kasakela males, and the infant she was carrying was injured. The males took the infant up into a tree, killed it, and began to eat it. Then, according to the observers, the males "seemed to think they were doing something wrong," and one of them carried the body 3.2 kilometers (2 miles) to Goodall's camp and left it on the lab porch. We hesitate to interpret the last part of that behavioral sequence. It was not an isolated event; infant deaths from similar attacks on female-offspring pairs by members of other groups occurred several other times at Gombe.

At about the time of these observations, the

Chimpanzees are the most abundant great apes outside of human beings—but we still outnumber them more than ten-thousand-fold. And their numbers are declining, whereas ours are still expanding, threatening both them and us.

Kasakela community split into northern (Kasakela) and southern (Kahama) communities. The Kasakela community included six prime males and two old ones; the Kahama group consisted of seven males, four in their prime reproductive years, one past prime, one old, and one adolescent. As is typical of territorial chimps, the Kasakela and Kahama groups patrolled their territories; compact groups of males and often females in estrus moved silently and purposefully through peripheral areas. During the early stages of fission in 1971, when Kahama males met Kasakela males, there were sometimes charging displays by the southerners, but generally interactions were peaceful. By 1974 encounters had become much more aggressive. Males from the Kasakela community began a series of violent forays toward the south, featuring a series of protracted, brutal attacks on members of the Kahama community. By 1977 the Kasakela males had wiped out the Kahama community, with all the healthy prime males killed (or missing and presumed dead). The restructuring of territories didn't end there; the destruction of the Kahama group removed a buffer between the Kasakela community and the Kalande community of nine males even farther south. The powerful Kalandes began to expand northward at the expense of the Kasakelas, who in turn moved farther north. Patrolling and tensions continued.

There is building evidence that such violent and aggressive behavior is widespread in chimp groups in the wild. But chimp cultures can differ from place to place, and the Gombe Stream chimps were living in an area of recently dramatically restricted habitat. In 1960, when Goodall first started her pioneering work at Gombe, the forest stretched unbroken 96 kilometers (60 miles) east from the shores of Lake Tanganyika. A decade later, it extended only 3.2 kilometers (2 miles) to the rift escarpment, a ridgeline that more or less paralleled the lake shore. Most of the forest beyond the ridge had been cleared for cultivation. That radical environmental change, conceivably leading to unusual crowding or resource shortages for the chimps, may have been a factor in generating the intergroup strife at Gombe.

For some decades wild chimps have been in decline. It is estimated that the sixty-year span from 1970 to 2030 will see a reduction of about 50 percent in the number of wild chimpanzees. On the plus side, chimpanzees are the most abundant of the great apes other than humans, with an estimated several hundred thousand individuals in equatorial Africa. Yet those numbers will only buffer them for a while. Their future is bleak because fast-growing human populations are converting their forests to agricultural land, cutting down the forests for wood, and mining for minerals and jewels in what was once pristine chimpanzee habitat.

Unchecked logging is bad enough for chimps, but even where the trees are not cut, the building of roads to get to harvestable tracts allows access by people. And some of the people are chimp hunters, who kill them for bushmeat or capture them for the pet trade or medical research. As with the gorillas, people transmit human diseases to the chimps, including Ebola. Add a road to a once-remote ancient forest, and you

Bonobos are our sexy living relatives. Once called pygmy chimpanzees, they are more slender and upright than standard chimps—that is, more like us.

will soon see the demise of wild chimpanzees and other bush wildlife.

Bonobos are, in some ways, more like us than are their phylogenetic twins, the chimps. For example, bonobos are more slender and more upright. Occurring in the Democratic Republic of the Congo, they are potentially more endangered because of their limited distribution, poaching, hunting, diseases, and habitat loss. The easy availability of modern firearms as a residue of civil strife means many people have the means to go bonobo hunting, and they do. An optimistic estimate of the bonobos' population size might be more than 50,000.

Although bonobos are not as well studied as chimpanzees, we do know that there are many aggressive interactions among males. Individuals hit, bite, slap, shove, grab, and otherwise abuse each other and use a variety of bluffing and charging displays. Aggression by a dominant male is countered by various forms of submissive reactions from lower-ranking males, including appeasing the aggressor by permitting him to mount. There is ample sexual competition among males, and dominant individuals appear to have more opportunities to mate, although lower-ranking males can be quite successful, perhaps because female bonobos have longer and more frequent periods of estrus than do chimpanzees. But there is so far no sign that they engage in intergroup murder as chimps have. Much strife is avoided in bonobo populations by their practice of relieving tensions with communal sex and female genital rubbing.

The "Man of the Forests"

Orangutans are our only great ape relatives found in Asia, on the islands of Borneo and Sumatra. In the local language, "orang" and "hutan" translate into "man of the forests." Orangutans are extremely intelligent, and an adventurous, somewhat lucky tourist can observe wild Bornean orangutans in the eastern part of the island, near the entrance of the Gomanton caves.

The caves are famous for their swallow nests, which collectors harvest by the thousands after the swallows' breeding ends. Some nests are taken from as many as 90 meters (295 feet) high on the ceiling of the cave, using elaborate bamboo ladders. Twice a year collectors risk their lives to gather the nests, which are sold for a good price as the basic ingredient for Chinese birds' nest soup. The caves are surrounded by a protected tropical forest that maintains a healthy population of orangutans, macaques, and other monkeys. Orangutans there are very calm and seem unafraid of humans; apparently they feel protected from poachers.

An hour from the caves is the Sepilok Orangutan Rehabilitation Centre, world famous for its conservation efforts to rehabilitate young orangutans, orphaned because poachers have killed their mothers and collected the juveniles to supply the pet trades. A few of the youngsters are intercepted and being rewilded at the center. These are the lucky ones; the staff train the young orangutans to survive and live on their own. Visiting the center can be a moving experience. Following a trail to the feeding platforms, a tourist can see many young orangutans and a few older animals. The orangutans roam freely in the tall tropical forest around the center.

In the same area is the Kinabatangan Valley, one of the last wetlands in Borneo, world-renowned for its elephants and abundant wildlife. Although large oil palm plantations, which displace the habitats of orangutans and other wildlife, have heavily impacted the valley, there are still large tracts of forest and wetlands. A boat ride to the mangroves will allow one to see orangutans and proboscis monkeys, another endangered species found only in Borneo. Proboscis monkeys live exclusively in swamps, riparian forests, and mangroves. The males, weighing about 24 kilograms (53 pounds), are twice as large as the females and have very large, conspicuous noses, for which they are named. Proboscis monkeys are endangered; their populations have declined by 50 percent in the past thirty years because of habitat loss and hunting. The

The Sumatran orangutan, endemic to the island and largely restricted to its northern tip, is another of our close relatives likely on its way out. Roads and timbering are fragmenting and destroying the forests that this arboreal ape depends on; they are hunted for the pet and zoo trades and killed where they compete with farmers for fruit.

Kinabatangan district in Sabah maintains one of the largest remaining populations.

Orangutans were once found on the Asian mainland from southern China to the Malay Peninsula, until about 12,000 years ago. By the seventeenth century, the species' distribution was confined to the islands of Sumatra and Borneo. Based on genetic evidence, a team of scientists proposed that the orangutans from the two islands, long considered a single species, were actually two different species. Most conservation agencies now consider the Borneo and Sumatra populations to be separate species. This is not an issue of exclusively scientific interest but one with important conservation implications. If they are two species, then the Sumatran orangutan is critically endangered because only about five thousand remain, a 60 percent decline in only two decades; they have a restricted geographic range in the province of Aceh in the northern part of the island. The Borneo orangutan is in better shape, because it inhabits a much larger area and its population size is around 40,000 to 50,000. The genetic differences also tell us that orangutans from one of the islands should not be introduced to the other island.

Even with its larger population, the Borneo orangutan is often difficult to see in the wild. But in their realm there are other marvels to enjoy. You might watch a red giant flying squirrel climb to the top of a tree on a forested ridge and then sail over the valley some 100 meters (330 feet) or more to disappear behind a line of trees. The general impression can be of a soaring furry garbage-can lid. The squirrel is not classified as endangered because of its rather wide distribution, but its populations are dwindling because of ubiquitous human-caused habitat destruction.

Orangutans are disappearing more quickly than are flying squirrels because their distribution is much narrower; their home ranges are larger; and much of their natural habitat is being converted to oil palm plantations, agricultural fields, and towns. Sumatra has lost practically all of its lowland forests, and Borneo is being deforested rapidly. In the Kinabatangan Valley and other regions in both Malaysian and Indonesian Borneo, plantations of palm oil (a rich cash crop) are quickly replacing natural forests and their wildlife. New oil palm plantations in national parks are destroying orangutans and populations of thousands, perhaps hundreds of thousands, of species of plants and animals. This is a horrendous tragedy, in which single or small family groups of orangutans are isolated in a forest patch of few trees, where they are abandoned until they die from starvation. This is part of the holocaust of nature, supported by the well-funded neo-colonialist propaganda organization World Growth, which despite its claims about conservation, is actually accelerating the impoverishment of the planet's life and beauty.

The Magnificent Elephants

Looking at the five-hundred-strong herd of African elephants in the tropical forests of the Chobe National Park in Botswana, one might wonder why such magnificent creatures are facing extinction, despite several long decades of international efforts to conserve them. Yet the number of African elephants today is less than half of what it was in 1979, when it was estimated at about 1.3 million. By then, they had already suffered greatly from hunting for their precious ivory—a big tusker can carry up to 100 kilograms (220 pounds)—as well as from habitat loss, the bushmeat trade, and diseases. Millions of elephants were slaughtered in the nineteenth and twentieth centuries for their tusks. At the peak of the ivory trade in the nineteenth century, up to 900,000 kilograms (2 million pounds) of ivory were sold to Western and Asian countries per year. Even so, the familiar savanna, or bush, elephant is still abundant enough to be an agricultural pest; a few individuals can destroy a poor farmer's entire year's crops in a few minutes.

Interestingly and surprisingly, the forest elephant, primarily a dweller in the rain forests of the Congo basin, has recently been declared a distinct, separate species. Like their better-known savanna-dwelling relatives, forest elephants have been decimated by ivory poachers. Their situation is precarious for many of the same reasons the bush elephant is threatened: growing human populations, habitat destruction, and (especially for the forest elephant) a demand for bushmeat to feed workers in expanding timber and mining

operations in central and west Africa. The elephants wander widely, playing critical roles in dispersing seeds and keeping forest paths and clearings open to the benefit of other animals. They pay no heed to reserve boundaries, areas being logged, or human settlements—behaviors that will not promote their survival unless serious efforts are made to protect them. The population of African forest elephants has plummeted dramatically in a few decades, from an estimated 1 million in 1950 to less than 17,000 now. In the past decade alone, the population fell more than 65 percent from 60,000 in 2002.

Today, despite international efforts to completely ban the ivory trade, an estimated 75,000 elephants are killed illegally every year for their tusks. Trophy hunt-

Asian elephants are the only elephants that can be easily tamed for work and entertainment. Like their African cousins, they are threatened by fragmentation of habitat—leading to hunting for ivory, meat, and leather—human-elephant conflicts over crop-raiding, and trapping for display in zoos as well as domestication for work in tourism and forestry. This type of idyllic scene in an Indian national park is becoming very rare.

ers legally kill a few thousand, and those animals bring revenue to many communities and countries. When properly organized, legal hunting of some species is potentially a powerful incentive for conservation, as is controlled hunting in South Africa. And, of course, the ecotourism value to many countries that elephants provide is immense and possibly sustainable.

It would be wonderful if in 2050 some of our descendants could still go to the Okavango Delta and enjoy the spectacles one can see today. In the delta it is possible to observe fifteen to twenty elephants playing in a small lake, submerging and springing up like breeching whales, thrashing the water with their trunks, and pushing young ones up onto their mothers' backs. Such a sight should be worth infinitely more than 50-yard-line seats at the Super Bowl!

The Asiatic elephant used to be distributed throughout most of Asia, from India and Nepal to the island of Borneo. Unfortunately, habitat destruction, poaching, and trade have severely reduced its populations. Now, they cling to existence in a few protected areas, widely scattered and often isolated from one another.

The Dwindling Species of Rhinos

The Ngorongoro crater is the world's largest intact volcanic caldera (cauldron-shaped depression), some 18 kilometers (11 miles) across and surrounded by a wall 400 to 600 meters (1,300 to 2,000 feet) high. Located south of the Serengeti National Park in Tanzania, the 260-square-kilometer (100-square-mile) crater floor, teeming with animals, is enclosed within the walls of the caldera of the extinct volcano. The ash fallout from this ancient volcano's eruption, forming a gradient of soil porosity as the bigger pieces rained down closest to the caldera, shaped the entire Serengeti ecosystem. The crater holds a sample of Africa's "big five game animals"—lions, elephants, black rhinos, cape buffaloes, and leopards. The crater still maintains a tiny number of the ill-tempered, solitary, two-horned black (or hook-lipped) rhinos, which make a living by eating the leaves of shrubs and other soft vegetation. These rhinos are now the jewels of the park.

As incredible as it may seem, in the late 1960s the total population of these magnificent rhinos was in the vicinity of 70,000 individuals. That was down from much larger populations that earlier occupied most of the eastern and southern portions of the continent. In the nineteenth and early twentieth centuries, black rhinos suffered heavily from poaching in connection with land clearing for agriculture. They were generally persecuted as dangerous animals, incompatible with human settlement. But the late twentieth-century surge toward extinction of two-horned black rhinos was caused by heavy hunting to obtain their horns, which were worth a great deal of money when used for traditional curved, carved rhino horn dagger handles (jambiya), especially in North Yemen, where new wealth from oil drove up prices. Another unhappy source of demand was (and is) in Asia, for phony medicines and for use as an aphrodisiac. In the late 1980s rhino horn was worth as much as $20,000/kilogram ($750 / troy ounce), well above

The Indian rhino looks like it is armored because of the way its heavy skin is folded. Unlike its big African relatives, it has only a single horn. Its prehensile lips are adapted to grasp the marsh grasses that comprise a major portion of its diet.

the price of gold at the time. A common sight on the African plains was a dead rhino, missing its two horns. Conservationists, trying to protect rhinos from poachers, have often tranquilized them and sawed off their horns.

One can hope that the advent of Viagra will reduce the demand for such butchery by lessening the value of rhino horns (and tiger penises). In Asia, particularly in China, the "little blue pill" may become a valued weapon in the battle to save the rhino and other charismatic species. A guess is that about 4,000 black rhinos exist today, thanks to conservation efforts that have brought them back from a low of about 2,500 in the early 1990s. But the threats to their existence in the wild remain, and only conservation vigilance will save them.

Other species of rhinos include the relatively tame and sociable, also two-horned, white rhinos, which are actually yellowish-brown or slate-gray in color. The white rhinos were misnamed, some say, because of confusion with their Africaans name, "wijd" rhino, referring to its wide mouth designed by evolution for grazing on grass. After elephants, the white rhino and the Indian rhino are probably the largest terrestrial mammals (although hippos may also sneak into that club). The white rhino is the least endangered rhino, with its stronghold in southern Africa. It numbers nearly 1,000 in zoos and perhaps 18,000 individuals overall. The white rhino has been persecuted for much the same reasons as its black cousins, but benefits from the relative ease with which conservationists can

The Sumatran, or hairy, rhinoceros is the smallest rhino. Except for the almost-extinct Javan rhino, it is the least known and the rarest. The Sumatran rhino's once rather extensive range in Southeast Asia has largely contracted to Sumatra, and fewer than one thousand are thought to exist today.

protect it because, unlike the solitary black, it forms herds that can be guarded in relatively small areas.

The magnificent Indian (or great one-horned) rhino once roamed over a huge range from India to China but is now confined to some protected areas in Nepal, India, and Bhutan and one small area in Pakistan. It is as big as the white rhino, and its deeply folded silver-gray skin gives it the appearance of being clad in armor plate. Perhaps three thousand still survive, mostly in Assam, despite intensive poaching for horns to use in attempts to cure impotency in Chinese men (more need for Viagra, which actually works).

The cruelty of large mammal poaching is exemplified by the fate of so many Indian rhinos. Shooting is by far the most common technique. Rifles and ammunition are often supplied by traders in rhinoceros horn to those employed to do the actual hunting. Pit-trapping is also used where the terrain is suitable, and the horn is simply sawed off the living animal. Sometimes, where power lines run through or near a reserve, an insulated rod is hung from the wires to a rhinoceros path and the animals are electrocuted. Add in the placement of rat poison or insecticides on salt licks frequented by rhinos, occasional spearing, and strangling the beasts with wire nooses, and you get a picture of the cruel way these impressive animals are being pushed toward extinction. The governments of Nepal and India have been struggling to protect these rhinos, with the help of the World Wide Fund for Nature (WWF); we can only wish them luck, for they face determined adversaries.

The other two remaining rhino species are in very deep trouble. The Javan rhinoceros is a close relative of the Indian rhino, and was once widespread in Southeast Asia but is now confined to Java. Until recently it also lived in Vietnam, but poachers reportedly killed the last surviving individual in 2010. It is now the rarest large mammal on Earth. It is so rare that one ecologist who had been studying Javan rhinos for a couple of years had never actually seen one, only its dung. A rhino did, however, walk through his camp when he was away for an hour! Fewer than fifty individuals survive, all in Ujung Kulon National Park in Java, guarded by rhino protection units. There are plans to attempt re-establishment at a second site in Indonesia. Javan rhinos usually perish in captivity; none are exhibited in zoos, and an attempted captive breeding program failed miserably. Poaching for horn and loss of habitat (especially due to wars) have reduced it to this sorry state, and only success of the conservation program stands between them and extinction.

The small Sumatran rhino has two horns like its African relatives and is now found in a fraction of its former range, only in Sumatra and Borneo. It is rather hairy and can survive relatively high in the mountains, but fewer than three hundred individuals remain. The reasons for its endangerment are basically the same as for the other rhinos, and it has proved very difficult to keep or breed in captivity. Recently there has been some limited success, but the captive breeding program remains controversial because some scientists claim the same money and effort put into preserving extant wild populations would be more successful in dragging the Sumatran rhino back from the brink of extinction.

The Devil's Disease

Moving away from the biggest threatened mammals, consider the fate of the Tasmanian devil. This "devil" is the largest surviving marsupial carnivore, a title it unfortunately inherited after the extinction of the thylacine (Tasmanian tiger). Even in the late 1980s nobody would have suspected that the then very abundant marsupial carnivore would soon be ravaged by a mysterious disease that would push it close to extinction. Tasmanian devils are hunters and scavengers, powerful, bulky animals that are able to deliver the strongest bite (relative to body mass) of any mammal. They are well known for their aggressive behavior; individuals viciously bite and scratch each other during encounters. The aggressive behavior becomes especially violent when Tasmanian devils fight over a carcass, much in the manner of hyenas.

In 1996 biologists found an individual with strange lesions on its face during a field survey. Many more sick or dead animals turned up in the following years.

With the extinction of the thylacine, the endangered Tasmanian devil is the largest carnivorous marsupial in existence.

By 2009 the species was classified as endangered because their population had plummeted more than 80 percent to fewer than 10,000 to 15,000 survivors. The disease is now known as the devil facial tumor disease (DFTD) and has been described as "an emerging infectious facial cancer" that is transmitted by infected cells passed directly between animals through the injuries they inflict on each other when fighting. This is an extremely rare mode of cancer transmission, almost unknown in nature. Because of the severity of the threat, conservation measures to save the Tasmanian devil from extinction include the destruction of sick individuals, the eradication of the disease where feasible in relatively confined areas such as peninsulas, the removal of healthy animals to a captive breeding program, and the establishment of populations on islands free of the disease.

Lords of the Night Skies

Representing about 20 percent of all wild mammal species, bats are the only mammals able to fly. Flying squirrels and flying lemurs are able to glide, but they can't actually power themselves through the sky. Bats are intriguing (and largely unappreciated) creatures that play essential roles in natural ecosystems and in maintaining human well-being. Insectivorous bats kill enormous numbers of insects that are crop pests or that transmit diseases to humans, as do mosquitoes. Huge bat colonies in northern Mexico and the southern United States, comprising up to 20 million individuals, can consume 40 tons of insects every night. The potential value of this ecosystem service

has been estimated at billions of dollars annually. But bats are also important because they pollinate many plants of economic or ecological value and disperse the seeds of many tropical trees. Without bats, many species of plants would become extinct for lack of pollination and/or seed dispersal. Unfortunately, diseases, impacts from introduced snakes, destruction of bat refuges and habitats, or direct attacks on colonies all threaten many bat species.

On Guam, the brown tree snake kills bats, as well as rails and other birds. A fierce predator, the snake has decimated three of Guam's species of bats. The Marianas fruit bat collapsed to fewer than fifty individuals, but has rebounded to around one thousand. Its survival is still precarious, and predation

The Mexican free-tailed bat is one of the most abundant and widespread mammals in the Americas. Seeing this picture, one can understand how bats can be major predators on insects—a vital element in natural pest control. Despite being present in huge colonies in some areas, this bat is declining. It is now feared that they may spread widely the deadly white-nose syndrome that has recently entered their range—as carriers even if the species does not itself prove highly vulnerable.

by the brown tree snake remains its main threat. The smaller little Marianas fruit bat fared even worse and is considered extinct. The third species, the Pacific sheath-tailed bat, was completely wiped out on Guam but survives on nearby islands.

A recent problem has emerged in the eastern United States and Canada, where a fungus disease called white-nose syndrome has been ravaging bat populations. Within five years of its first appearance near Albany, New York, in 2006, the disease had killed more than a million bats of several insectivorous species, from North Carolina to New Brunswick and west to Missouri and Oklahoma. The victims are most vulnerable in winter. While the bats are hibernating in close-packed masses of hundreds to thousands in caves or mines, conditions seem especially favorable for the fungus. Of the six or more bat species affected, the one most devastated has been the little brown bat, normally the most common in the region. Mortality in little brown bat colonies has been more than 90 percent within one or two years of first infection. The disease also exists in Europe, but European bats appear to be much more resistant to it. Whether or not the North American infection was a transfer of some sort from Europe is being debated, though many researchers suspect a cave explorer accidentally brought it to America.

White-nose syndrome first appears as pale fibers over the bats' faces—hence the name. It spreads to other parts of the body, invading and "digesting" the skin of the wings. During hibernation an infected bat may awaken, needing food because the fungus has robbed its energy. It may leave the cave in search of food to replenish stored fat, but with no insects available in winter, the bat usually starves or freezes.

If scientists' estimates are correct, bats provide billions of dollars worth of pest-control services each year to agriculture alone. They may also help to control several insect-borne diseases such as viral encephalitis (including West Nile virus) in the United States, as well as malaria and dengue fever, which are likely to become more common in southern states as the climate warms.

Scientists in Canada and the United States are working to find ways to defeat white-nose syndrome, including genetic studies to discover the reasons for its lethality and testing fungicides against it. There are also efforts to make the bats' summer habitats more hospitable, which seem to have stabilized some of them, and helped the decimated colonies. Other scientists are studying aspects of bat behavior that could reduce their vulnerability and looking for signs of genetic resistance in the survivors, which might then be used to rebuild the colonies.

The Oddest Mammal

On a lovely warm evening in November, a tranquil stream just southwest of Sydney, Australia, seems quiet. Yet beneath the water's surface, one of the weirdest of all mammals, the platypus, is busy foraging for a meal. This small duck-billed, beaver-tailed, egg-laying, semi-aquatic animal is unusual, to say the least. Having a duck-like bill should be enough to make us wonder if they are mammals at all, but there is more.

The males wield poisonous spurs on the back of their hind feet, capable of inflicting extremely intense pain. The females lay eggs. The feet of both sexes resemble those found on waterfowl. Their bills are equipped with electrical sensors.

When platypuses forage along the bottoms of their quiet, shaded streams, they keep their eyes, nostrils, and ears closed. A sensory system in their bills responds to changes in pressure and in electrical fields, enabling them to find their small invertebrate prey. When resting or producing young, they hang out in burrows in or close to stream banks. This extraordinary animal is an excellent example of an animal with the most ordinary conservation status on Earth. It is a widespread species and not in immediate danger of extinction. But as is the case for so many mammals and birds, it is suffering from the gradual erosion in numbers that human activities often cause.

Platypuses are still found in much of eastern Australia. Aboriginal Australians hunted them for food by spearing them when they surfaced, but probably didn't seriously affect their populations. Later, European settlers hunted them commercially for their pelts, killing many thousands before the practice was banned. Commercial net fishing in streams once sometimes decimated platypus populations, but this method was also banned. So, why are they in trouble, like so many other forms of wildlife? It is the cumulative effect of many seemingly indirect human actions. A list of the impacts reveals the problem. Although net fishing is illegal, it continues to some degree. Additionally, declining water quality of streams in urban areas can reduce platypus population sizes or exclude these animals entirely. Platypuses can survive icy conditions but are sensitive to warm water, and Australia's climate is certainly warming. The recent introduction of red foxes in Tasmania may well result in predation on platypuses when they travel over land between streams. Finally, the invasion of Australia by common carp is radically altering the fauna in the continent's streams, and the result is almost certainly going to be negative for platypuses. In short, platypuses are, like the vast majority of mammals around the world, suffering a "death of a thousand cuts" at the hands and machines of *Homo sapiens*. They are examples of an animal beset by the not-so-sexy, but very real, set of conditions that increasingly plague the overpopulated (by humans) Earth.

Extinction of a Conservation Icon

Some large mammals are gone from the wild, but still persist in captivity. One classic example has become a conservation icon: the slender-necked reddish-tan-to-gray Père David's deer, which, unlike most of its relatives, is quite at home in water—sometimes seeming more like a seal than a deer. We know about this deer thanks to a determined French naturalist, Abbé Armand David, who discovered it in 1865. He didn't see it while adventuring in the swamps of the Chihli plains in northeastern and east-central China, where it was once abundant. He could have been as many as three thousand years too late for that. The swamps had been drained during the Shang dynasty (1766–1122 BC), and the deer ceased to exist in the wild in that area. No, he risked his neck to spy it over the wall of a closely guarded imperial hunting park near Beijing.

Père David had no trouble recognizing the deer as different from other deer species; they had unique forked hind prongs on their antlers, long tails, and broad hooves well suited for their original boggy habitat. Père David's deer was once quite widespread, occurring in the late Pleistocene (ten thousand or so years ago) in Korea and Japan as well as China.

The large hunting park herd that Abbé David spied was eventually destroyed in the early twentieth century by a flood, starving peasants, and foreign troops putting down the Boxer Rebellion. The species was saved from extinction largely because of the efforts of the Duke of Bedford, who used a few illegally obtained deer to found a captive herd, which he protected assiduously through two world wars. Individuals from the duke's herd were later dispersed to captive facilities in many parts of the world. Père David's deer has now been successfully reintroduced in several areas in China, including the Beijing Milu Park and the Dafeng Milu Nature Reserve. At present, the total population numbers nearly one thousand.

The Antelope of the Steppes

The steppes of Central Asia, in Russia, Kazakhstan, and Mongolia, formerly hosted huge herds of the saiga, an extraordinary antelope, exquisitely adapted to those landscapes. Millions of saigas, a nomad species that once was one of the most numerous large herbivores on Earth, lived in semi-arid rangelands where they had to cope with extremely cold winters and hot summers. These antelopes have a peculiar nose that allows them to filter sand during the hot and arid summers.

Saigas are a classic story of the brutality of human persecution. Hunted for their horns, meat, and pelts, their population decreased to fewer than 30,000 by 1950. The Russian government launched an ambitious conservation program that was very successful, and by 1990 the saiga population had increased to more than 1,200,000 animals. Unfortunately, the crumbling of the Soviet Union had as a collateral consequence the collapse of conservation efforts and the precipitous decline of saiga populations. By 2000 an estimated 178,000 saigas remained; now, there are no more than 50,000, and their population is still decreasing, mainly because of poaching and diseases. In a single week in 2010 more than 12,000 saigas died from an outbreak of pasteurellosis, a bacterial disease.

Hornaday of the Smithsonian

The American bison was one of the most abundant large mammals on Earth when the first Europeans arrived in the Americas. It is estimated that 30 to 60 million bison roamed the vast plains and woodlands from Alaska and Canada to northern Mexico. By 1880 they were nearly extinct and might well have disappeared without the intervention of one man, William T. Hornaday of the then-fledgling Smithsonian Institution. In 1889 Hornaday wrote of earlier times, when the bison were abundant: "They were so

numerous they frequently stopped boats in the rivers, threatened to overwhelm travelers on the plains, and in later years derailed trains and cars."

In those days, a single herd in the Arkansas River valley was estimated to have 4 million animals. There are many similar accounts of the incredible abundance of bison in the early 1800s. Indeed, when you read the old accounts, it is almost impossible to believe that they became nearly extinct eight decades later. They were relentlessly hunted for their hides and meat. The slaughter was brutal. The North American Fur Company, one of the leading fur dealers at the time, sold 67,000 skins in 1840, 110,000 in 1843, and more

The American bison was saved in Yellowstone National Park. Presently, the bison population in the park numbers roughly four thousand. The Lamar Valley, in the northern part of the park, concentrates a large part of the population, which can be easily seen by tourists.

than 250,000 in each of the following two years. Then the supply was dwindling, and by 1860 very few skins were traded.

It is claimed that "Buffalo Bill" Cody bagged four thousand bison in a single year! By 1870, most large herds were history. Hornaday warned constantly of the possible extinction of the bison if no protection were granted to the species. One of the last big herds found refuge in the Texas panhandle. Estimated at 3 million bison in 1871, the so-called southern herd was exterminated for political reasons. U.S. Army Generals William T. Sherman and Philip Sheridan viewed the eradication of bison as "the critical line of attack" in the war against the Plains Indians. According to the historian David Smits, Sherman wrote to his friend Sheridan in 1868: "as long as Buffalo are up on the Republican the Indians will go there. I think it would be wise ... [to have] a Grand Buffalo hunt, and make one grand sweep of them all."

When some U.S. congressmen were considering granting the southern herd conservation status, General Sheridan convinced Congress that exterminating the bison was the only way to "pacify" the Plains Indians. Between 1872 and 1874, almost half a million skins were freighted east by the Santa Fe Railroad. In 2 months (December and January) of the winter of 1877–1878, more than 100,000 bison were killed, and 5 years later only scattered groups of bison were left. The great southern herd had been destroyed.

The "tongue story" illustrates the zeitgeist of 1800s North America. Bison were so numerous that many times hunters took only the tongue, which was considered a delicacy, and left thousands of carcasses to rot. Smits tells of one hunting trip: "In three days they returned with a wagon filled with more tongues than were ordered. To kill over 144 buffalo, animals that could weigh over 2000 pounds each, solely for their tongues, which weighed an average of two pounds apiece, was perfectly justifiable to those frontiersmen who believed the herds were expendable."

Because of the alarm calls of Hornaday and others, extinction was prevented just in the nick of time. Late in the nineteenth century, only 325 bison remained in all of the United States. Today there are around 3,500 bison in Yellowstone National Park alone, all descendants of approximately 25 bison that were protected when the park was created in 1872. Yellowstone, one of the wonders of the world, maintains an impressive diversity of large mammals, and it is easy to have magnificent views of bison, elk, and pronghorn antelope in the Lamar Valley, a sight that seems drawn from past centuries.

Over all of North America more than half a million bison now survive, essentially in captivity, on private ranches in Canada, the United States, and Mexico. In comparison, there are only about 20,000 wild bison left in conservation herds (i.e., not for commercial sale) in parks and on some private ranches. Most are on landscapes of less than 809 hectares (2,000 acres), although some are in large remote wilderness areas.

The view of great herds we see in Africa today are not that different from how it once was in America. We lost our natural wonder—millions of thundering, migrating bison—and retain but a shadow of the spectacle that once awed so many. Perhaps the lesson of the bison will instill a desire in humanity to preserve the vast herds that remain elsewhere. Let us hope the spirit of Hornaday will re-emerge to save the wild mammal herds.

The polar bear, symbol of the Arctic, is a giant carnivore, closely related to the brown bear. It is now threatened by tiny molecules of carbon dioxide added to the atmosphere that are warming the planet and melting the ice that is a vital portion of its habitat.

8 WHY IT ALL MATTERS

When Rachel Carson launched the modern environmental movement in 1962 with her scientifically accurate and beautifully written book *Silent Spring*, she said of pesticides, "These sprays, dusts, and aerosols are now applied almost universally to farms, gardens, forests, and homes—non-selective chemicals that have the power to kill every insect, the 'good' and the 'bad,' to steal the songs of birds, and the leaping of fish in the streams."

Carson's book inspired the public and drove politicians to act. Some dangerous pesticides such as DDT (dichlorodiphenyltrichloroethane) were eventually banned in most of the world, and the use of others was better controlled. As a result, the disappearing populations of some of the most interesting North American birds, such as the bald eagle and peregrine falcon, began to rebound. Banning of DDT didn't make it disappear entirely. It is with us today; traces are found in environments and organisms (including people) around the world. The poison is still killing birds in St. Louis, Michigan, site of a Velsicol Corporation plant that manufactured the pesticide until 1963 and left behind a toxic stew containing DDT in soil and groundwater. But most of the dangers confronting nature were not removed by the environmental laws; human impacts on the environment have turned out to be much more complex and far-reaching than we once thought. The measures that Carson inspired have been more than countered by other factors in recent decades. The celebrations of the 1970s were premature, for the trends in North America and elsewhere point to an overall downward curve not just for birds but also for the populations of most other wild species.

Such downward trends will have repercussions for humanity in a multitude of ways. All animals, including birds and mammals, are intricately involved in the processes and interactions of other organisms in natural ecosystems. For example, land animals pollinate flowers; disperse seeds; transmit disease-causing organisms; and consume the roots, stems, seeds, foliage, and flowers of plants. Many prey on other animals, including those that attack our crops. Most of their interactions with human beings are positive.

Aside from the ethical tragedy of diminishing the stock of life on Earth, the most serious consequences of the declines and disappearances of animal species are the losses of critical natural services that emerge from the global system—losses that have important economic, social, and political implications for humanity. Nature's services (ecosystem services) are the benefits that we humans obtain free from the structure and functioning of ecosystems, which are essential for maintaining life on Earth. Put simply, populations of wild species and their activities are vital for maintaining the entire global system on which humanity depends.

Plants and Their Pollinators

Many animals and plants have mutually beneficial relationships. Plants provide food for animals; many animals disperse the plants' seeds and are thus intricately involved in the sex lives of the plants. Pollination is the process by which pollen is transferred from the male reproductive stage of a plant to the female reproductive organ, usually of another plant individual, to form seeds. We have all heard the buzzing of myriads of European honeybees and seen them actively seeking the nectar and pollen of flowers in gardens. While moving from plant to plant, the bees get coated with pollen (actually tiny male plants) and transfer the grains between flowers. Indeed, nectar is a "reward" the plants have evolved to attract animal pollinators.

Not surprisingly, the number of plants pollinated by animals is staggering, and the services provided by pollinators have huge direct effects on the human economy. At least eighty-seven of the leading global food crops are dependent on pollination by animals, including insects, birds, and mammals. Up to 70 percent of all the vegetables and fruits in our diet, such as apples, pears, mangoes, bananas, guavas, squash, almonds, chocolate, and passion fruit, among many more, are pollinated by animals.

The pollination process is fascinating, with extraordinary specializations of the plants and animals linked by this relationship. Besides bees and other insects, enormous numbers of bats and birds help pollinate the vast diversity of plants, especially in the tropics. For example, bats pollinate the agave plant from which tequila is brewed. Without bats, there would be no agave and no tequila. Bat-pollinated flowers generally are open at night. They are relatively big, with wide mouths, pale coloration, and sturdy petals, and produce strong fruity fragrances that often are described as musky or fermented. Some plants, such as the balsa tree, produce so much nectar that they attract many animals to their flowers, including mammals such as capuchin monkeys. Nectar-feeding bats have several specializations related to their feeding habits, such as long noses, long tongues, and the ability to hover when approaching the flowers.

On cool February nights, thousands of southern long-nosed bats visit the flowers of cazahuate trees in the tropical dry forests of western Mexico. The bats approach the flowers flapping their wings, stick their heads in the flowers, and lick the nectar. In the process, their heads become covered with pollen grains, which will pollinate flowers visited afterward. These bats migrate every year from the warm tropical forests in western Mexico to southern Arizona, following the rhythms of flowering, to form large maternity colonies in a very few caves. There, hundreds of thousands of southern long-nosed bats give birth more or less simultaneously, within two to three weeks. Once the young have become juveniles and are able to fly, they return to Mexico to provide pollination for the cazahuates.

Birds play an even more important role as pollinators, since more plants depend on them; some 250 species of plants in Australia alone depend on more than 100 species of birds for pollination. In North America hummingbirds, which have the fastest wing beats—up to an astonishing 20 beats per second—specialize in feeding on nectar and pollen, pollinating many species of plants as they feed. Hawaii is famous for its nectar-feeding birds, but many, unfortunately, are already extinct or endangered. Many nectar-feeding birds have long beaks and tongues specialized to

A Mexican long-nosed bat performs a pollination service. Bat-pollinated flowers tend to be big, showy, strong smelling, and, curiously enough, open at night!

extract the nectar out of the long tubular flowers of the plants they visit and pollinate. Among these are the hummingbirds of the Western Hemisphere, the sunbirds of tropical Eurasia and Australia, and honeyeaters in Australia and the Pacific Islands.

Many plants depend on very few or a single species to pollinate them or disperse their seeds, and many animals depend on only a few plant species as food. Because of their often tight relationships with particular plant species, hummingbirds and other specialized birds can be especially vulnerable to extinction, and so can plants that depend on specialized pollinators or seed dispersers. In some ways, these species are "trapped" in such relationships, so the fate of one hangs on the fate of the other. If many such links are broken, as often happens due to clearing of habitat for agriculture, cattle-grazing, or other human activities, an entire local ecosystem may collapse, losing many of the attributes that are valued by humanity.

Remember the brown tree snake introduced to Guam that caused the loss of numerous forest birds and some bat species? Among the species going extinct were important pollinators and seed dispersers, whose disappearance had downstream repercussions for the ecosystem. The number of birds pollinating plants in Guam was so drastically reduced that one study found that, between February and May 2005, no birds were observed visiting two plant species, the large-leafed mangrove and the coral tree. These two plants depend on birds for pollination and thus are unable to reproduce in their absence. Seed dispersal also seems to have been compromised because of the

loss of the birds and bats that once performed that function. Indeed, significantly fewer saplings of the large-leafed mangrove can be seen on Guam than on the nearby tree snake–free island of Saipan, where fruit-eating birds still persist.

On the Hawaiian Islands, at least sixty species of birds, including the akohekohe and the iiwi, have been lost or are experiencing severe population declines. The downstream damage? It is suspected that the loss of these avian pollinators has caused the elimination of more than thirty native Hawaiian bellflower species. Another well-studied example concerns a New Zealand shrub whose plant density, seed production, and pollination have been reduced in areas where bird pollinators such as the bellbird have been extirpated. Seed production per flower has been reduced by as much as 84 percent and the number of young plants by 55 percent in the absence of the birds.

The services of mammals, birds, bees, and other pollinators in producing commercial crops are valued at billions of dollars every year. Animal pollinators are fundamental to maintaining and increasing the productivity of cash crops such as coffee, peaches, and apples. Unfortunately, no economic estimates have been made of the potential costs of losing bees, birds, or bats and other mammals as pollinators to natural ecosystems worldwide.

The loss of birds such as the akohekohe (*top*) and the iiwi (*bottom*), endemic to the Hawaiian archipelago, has a large impact on the ecosystem because they pollinate many endemic species of plants. The akohekohe is now restricted to Maui and is endangered, but the iiwi is fortunately common throughout the archipelago.

Dispersing the Seeds

Like pollination, seed dispersal by animals is a key ecological process. Plants rely on wind, water, and animals to spread their seeds—that is, to transport them

The oriental pied hornbill is the most frequently seen hornbill in Asia and seems for the moment globally secure, unlike other Asian hornbills, which are generally declining. Nonetheless, the oriental pied hornbill has been virtually wiped out in southern China—in some areas it is captured for the pet trade, and its casques (helmet-like structures on the bill) are commonly made into souvenirs.

from their source to near or distant suitable habitats where they can successfully germinate. In many places, animals are the principal transporters. Roughly 90 percent of tropical tree species rely on mammals and birds for that service, producing large, oil-rich fruits to attract and feed them. In some areas, such as African tropical forests, avian frugivores (fruit-eaters) are the most important agents of seed dispersal, spreading more seeds than do other frugivores such as primates.

Even so, monkeys, apes, and even elephants play significant roles in dispersing seeds. The long coevolutionary history of plants and their bird and mammal dispersers has produced spectacular specializations to strengthen the relationship. If the bird or mammal is active in the daytime, the fruits have bright colors; in contrast, fruits with seeds spread by nocturnal bats have relatively large sizes and strong odors. In some cases the seeds cannot germinate until they have passed through the guts of their dispersers. In that process, the seeds are scarified, which means that their protective coat is thinned, so germination can occur after the animal defecates the seeds. The thinning is essential to enable the seeds to absorb water from their environment and initiate the germination process.

Reduction in the numbers of fruit-eating birds and mammals can, by hampering seed dispersal, lead to the rarity of some plants and have negative consequences for forest regeneration in particular and ecosystem function in general.

Frugivorous birds such as hornbills and pigeons are heavily hunted in many countries, such as the Philippines and Indonesia. While visiting various national parks in Indonesia, Navjot Sodhi witnessed poachers using blowpipes to kill birds that were visiting large trees to feast on their fruits. While visiting the only primary forest in Singapore, the Bukit Timah Nature Reserve, Sodhi found fallen fruits and seedlings directly under many large trees where, shaded by their parents, they couldn't thrive, an indication that their avian dispersers had disappeared from that majestic forest. With their reproduction so curtailed, most of those tree species too would eventually disappear from the reserve.

Because of widespread fragmentation of tropical forests in general, many such ecological interactions remain at risk, and many doubtless have already been lost. In the tropical forests of the New World, birds such as toucans and curassows are among the most important seed dispersers, but hunting is depleting their populations. The impacts of such losses can be far-reaching. For example, about one-third of the tree species in Brazil's Atlantic Forest may go extinct in the future, due to the loss of avian dispersers. Interestingly, some of the trees are evolving responses to the loss of dispersers. A keystone palm species produced large seeds whose dispersal required big birds capable of opening their beaks wide (large-gaped) to disperse them. As populations of large-gaped birds, toucans and big cotingas, were decimated, palms in areas where the birds once were abundant rapidly evolved smaller seeds. Such seeds are less likely to produce successful trees, however, and may make the palms more susceptible to local extinction. Generally speaking, the extirpation of larger bird species from tropical forests clearly will have dramatic, and often deleterious, effects on their tree communities.

Many mammals are also important seed dispersers that have intricate relationships with the plants whose seeds they spread. For example, gibbons, muntjac deer, and sambar deer, the only seed dispersers of a forest canopy tree known as the Nepali hog plum, are all extensively hunted in Thailand. Heavily hunted Thai parks have exhibited lower seed dispersal and fewer seedlings of the Nepali hog plum. Frugivorous mammals in these parks are important for the trees' survival. They spread the seeds away from the parent tree into sunny forest gaps, where the chances of seed germination and sapling survival are relatively high because of a lower concentration of seed predators and the availability of sunlight. This tree species may eventually become extinct in areas where these mammals have been extirpated.

Many forests are only partially emptied of bird and mammal frugivores. The ecosystems of these "half-empty forests" contain reduced but still viable frugivore populations. In these forests, only trees bearing preferred fruits may receive visits, leaving less-favored plant species without dispersal agents.

Therefore, even a partial reduction of populations of fruit-eating mammals and birds may exert a significant negative impact on the numerous plant species that depend on them.

Frugivores can also play essential roles in the countryside, human-dominated landscapes where forest patches may be expanding or contracting, depending on the patterns of human activity. However, it is usually unclear whether sufficient numbers of frugivorous birds are present in human-dominated landscapes (e.g., agricultural fields, urban areas, or suburban areas) to adequately disperse the seeds from the fruits.

Fruit bats also are essential for maintaining many Old World tropical and subtropical plant communities, but populations of these flying mammals have been declining dramatically because of habitat loss and hunting. In Ghana an estimated 140,000 large fruit bats are slaughtered each year for food. On tropical Pacific islands several bat species, such as the Guam flying fox and the large Palau flying fox, are believed to have vanished because of hunting and habitat loss. The decline of large fruit bats, valued as human food, may seriously limit the long-distance dispersal of large-sized seeds in much of Asia. Since fruit bats are more capable than other animals of taking seeds into degraded landscapes, the prospects for forest recovery in these areas are poor without healthy fruit bat populations. Because very few plant species are known to be exclusively pollinated by bats in Asia, the effects of a loss of frugivorous bats on pollination may be minimal. Nonetheless, as an indication of the impressive economic benefits of fruit bat populations, at least 450 economically significant products, including fruits, timber, medicines, tannins, and dyes, are derived from plants whose pollen and/or seeds are spread by them.

The decline of fruit bats and loss of their services may depress human well-being and economic aspirations in many areas. Flushing fruit bats from forests is known to jeopardize human health and enterprises. For example, catastrophic fires in 1997–1998 in Indonesia burned at least 5 million hectares (12 million acres) and decimated many wildlife populations. The fires were a consequence of poor land-use practices, such as over-intensive slash-and-burn agriculture, and exacerbated by drought conditions due to a severe El Niño event. The fires led to the failure of many forest trees to flower and fruit.

The smoke and haze produced by these enormous fires blanketed all of Southeast Asia for many weeks and caused at least U.S. $4 billion in losses to local economies, primarily because of declines in tourism. It is estimated that in the worst-hit areas, the smoke effects were equal to smoking four packs of cigarettes a day! During the peak of the smoke haze period, when Navjot Sodhi was sampling birds in Borneo, he suffered from watering eyes, difficulty breathing, and headaches. The damage to human health, especially to the more vulnerable members of the society—the poor, children, and old people—was severe. More than 200,000 people were hospitalized, and there likely will be long-term effects on many millions more.

The indirect biodiversity-related dangers such fires pose to human well-being are vividly illustrated by the

Fruit-eating bats, like this Jamaican fruit-eating bat, play a critical role in forest ecology by dispersing the seeds of important trees.

spread of the Nipah virus in Southeast Asia, which is intricately linked to forest fires in the region. Lacking sufficient food after the fires, fruit bats in Malaysia started visiting fruiting trees near large pig farms. The bats apparently passed the deadly Nipah virus to the pigs, which then transmitted it to humans. As a result, 105 people perished and more than 1 million pigs had to be slaughtered. This sad story illustrates how human destruction of nature can in turn injure and kill people.

Keystone Species

Keystone species are defined as those species whose positive impact on the environment is much larger than would be predicted solely by their abundance. The disappearance of keystone species can provide remarkable insights into the usually unforeseen negative effects of human activities. The importance of keystone species was first recognized a half-century ago in a classic study of marine invertebrates. That research showed that the presence of a starfish had disproportionate effects on the structure of an intertidal food web on the coast of Washington State. More recently, numerous scientific studies have demonstrated that some species, such as beavers and the prairie dogs we discussed earlier, are both keystone and ecosystem engineer species—those whose activities change both the species interactions and the physical structure of the whole ecosystem.

Beavers have the singular habit of constructing dams with bases of mud and stones and superstructures of brush and poles, which in turn are plastered with more mud and rocks, in the streams where they live. The dams create flooded areas where the beavers build dome-shaped stick-and-mud lodges to dwell in. Beaver dams are a common sight in forested areas in Canada and the United States. Besides providing a home for beavers, the dams create habitat for fishes, salamanders, frogs, invertebrates, and plants, enhancing the local biodiversity. In some areas the dams serve as critical nurseries for salmon; when beavers have been extirpated, salmon runs have dwindled. In general, when beavers disappear, the abundances of many of the associated species sharply decline, and in many cases they become locally extinct. Many beaver dams play important roles through such functions as flood control, breakdown of toxic substances, creation of wetlands and meadows, and removal of excess nutrients that can cause over-fertilization of streams. Beaver dams also may contribute to an important stage in the nitrogen cycle, one of several essential geochemical cycles that humanity is seriously perturbing.

Of course, from a human viewpoint, beavers are sometimes considered pests—they may, for example, flood roads or fields and cause considerable inconvenience. Although they once were threatened with near-global extinction through overhunting for their valuable pelts, they are now abundant over much of their range and only threatened occasionally with local extirpation. Nevertheless, beavers are still heavily hunted in the United States and Canada, where more than 40,000 animals are legally taken each year for the fur trade.

Sadly, those other ecosystem engineers, the prairie dogs, have had no such renaissance. Despite an increasingly detailed scientific literature on the value of prairie dogs as a keystone species fundamental to maintaining the native plants and animals of grasslands, the U.S. government still subsidizes killing them. Paradoxically, besides spending millions of dollars to eradicate the prairie dogs, the same government is spending millions of dollars to try to save threatened and endangered species such as kit foxes, ferruginous hawks, burrowing owls, and mountain plovers, all of which depend on the habitat created by prairie dogs to maintain viable populations.

Acacias, Ants, and Herbivores

A very interesting example, recently described, of the downstream effects of declines and losses of mammal populations is that of the consequences of the loss of large herbivores. As shown more than half a century ago by the notable tropical biologist Daniel Janzen, many species of ants and acacia plants have intricate cooperative (mutualistic) relationships. It is a common sight to see ants crawling on trees throughout the tropics. Some of the trees provide housing for the ants, and in certain instances, secrete carbohydrate-rich nectar from glands at the bases of leaves to feed the ants. In turn, the ants protect the plants from insect and mammalian herbivores. For example, the whistling thorn tree in East Africa produces swollen hollow thorns to house the ants that provide protec-

The tusks of the African elephant, sadly, are worth larger and larger fortunes on the black market that sells ivory to increasingly wealthy Chinese people. The disappearance of this elephant has had surprising unforeseen consequences. For example, in the savannas of Kenya, bacterial diseases transmitted to humans by fleas are more frequent where elephants and other herbivores have disappeared, because of the increase in the number of mice.

tion. But what happens to the ant-plant interaction when herbivores go missing?

In Kenya, whistling thorn trees are tended by four species of ants, with varying degrees of reliance on and benefits to the trees. The four ant species compete for exclusive use of an individual tree. Under natural conditions, the most common ant species, the cocktail ant, appears to be a classic mutualist; it uses the food and housing provided by the tree and patrols aggressively, repelling all invaders, including elephants (they don't like swarms of stinging ants rushing up their noses any more than we would!). The next most common species, in contrast, appears at best indifferent and at worst actively antagonistic toward the tree. This species does not use the housing provided by the tree; rather, unlike the other ants, it allows long-horned beetles to bore deep holes in the tree in which the ants nest. Neither does it rely on the tree's food offerings, opting to forage off the tree. Finally, this ant can hardly be bothered to defend the tree from attackers, and it is an unfaithful partner, frequently abandoning trees to occupy others nearby. The other two species of ants behave in ways that reduce the possibilities of a hostile takeover by one of the other species, trimming off branches that could be used for invasion from neighbor trees, destroying resources (e.g., nectaries) other species depend on, and so on.

A key finding about this complex web of interaction was that its stability depends on the presence of large herbivores. When scientists fenced large mammals out of an area of whistling thorn trees, the ant-plant mutualism began to unravel. After large mammals had been excluded for ten years, the plants evolved to invest less in providing nectar and housing for the ants. This weakened the colonies of the mutualist ants, which relied heavily on the rewards, and forced them to become more antagonistic to the tree by importing sap-sucking scale insects to produce honeydew as an alternative food source. Moreover, the weakened colonies of the mutualist ant species lost further ground to the unfaithful, cavity-nesting species, which became numerically dominant. The increasing dominance of the cavity-nesting species led to more long-horned beetle attacks, with attendant increases in tree mortality.

Removing large browsing mammals clearly was not as beneficial for whistling thorn trees as one might have predicted. The local extinction of megaherbivore populations has repercussions for ant-defended acacias and possibly for many other close, mutually beneficial interactions among organisms. This is a complicated and still only partially understood example. A better example might be the complex relationship between elephants and the habitats they create for smaller ungulates and wildlife and how their removal changes entire landscapes.

The Safest No-Cost Pest Removal

The control of herbivorous insects by birds is likely of great value in both managed and natural forest settings. Generally, insect herbivores inflict substantial damage in the canopies and understories of forests that lack insect-eating birds. Insectivorous understory birds (those that live and forage below the forest canopy) decline with increased disturbance and fragmentation of a tropical forest. The consequences of the loss of these birds on tropical forest productivity are potentially significant for other species and need to be carefully examined. In general, insectivorous birds are among the most sensitive of all bird groups to changes in their habitat.

Population declines and extinctions of insectivorous birds carry adverse implications for both wild and cultivated plants, including agricultural crops. Pests, including herbivorous insects, consume between 25 and 50 percent of the crops produced annually worldwide. Insects are estimated to cause most of the damage on cotton and about 10 to 20 percent of the losses of other major crops. Every year about $25 billion is spent on pesticides to control insect pests in the United States alone. The pesticides often pose threats to human health and cause environmental damage in many ways. In addition, they usually quickly become ineffective because pests easily evolve resistance to them, with classic examples involving overuse of DDT. Worse yet, pesticides are frequently more deadly to the predacious insects that normally limit the population sizes of plant-eating insects and, in decimating them, often promote previously harmless herbivores to pest status.

Changing public attitudes and consumer trends are now increasingly endorsing organic farming, in which other techniques for pest control are substituted for chemical pesticides. Organic farming should put a premium on pest control by birds, but it is difficult to predict the magnitude of their potential contribution. For example, it is very likely that insectivorous forest birds visiting crop fields adjacent to tropical forests help suppress agricultural pests, but this question has been poorly studied. More research is needed to determine whether the presence of insectivorous birds broadly indicates healthy agricultural ecosystems. In the long run, their free pest control services will likely gain prominence as the world's toxification problems become ever more prominent and potentially lead to greater restrictions on the use of pesticides.

Some studies already reveal that insectivorous birds do play an important role in controlling outbreaks of herbivorous insects in agroforests (stands of trees managed for production of certain crops). Coffee is a commercially important small understory shrub that traditionally has been grown on shaded plantations—that is, among forest trees. However, sun-exposed coffee farms, devoid of forest trees, have sprung up across the tropics in order to produce faster, higher yields. Yet shaded coffee plantations harbor a significantly greater diversity and abundance of birds than do sun plantations. It might be expected that large numbers of insectivorous birds would control coffee pests effectively on shaded plantations. Indeed, when ecologists simulated a pest outbreak by placing caterpillars on coffee plantations in Chiapas, Mexico, birds quickly came to the rescue, consuming the caterpillars. But this prompt pest predation response occurred only on a farm with high floristic diversity and not on a sun coffee farm. In some other studies, the amount of insect consumption by insectivorous birds did not differ between shade and sun coffee plantations. While more research is needed to adequately assess how land management practices affect the pest-control services provided by birds, it is nonetheless clear that preserving bird diversity can be crucial for maintaining control of agricultural insect pests. Birds appear to reduce insect abundance most in places where the diversity of bird species is highest.

Insectivorous bats are also important for consuming insects in coffee plantations. Researchers have found that bats reduced populations of arthropods (insects, spiders, mites, etc.) by 84 percent in a shaded coffee plantation in Mexico. Other studies indicate that insectivorous bats may remove pest insects even more efficiently than birds do. Consequently, the decline in bat populations worldwide will certainly jeopardize this vital ecosystem service, not only for coffee growers, but for other agriculturalists as well.

Predation on insects may translate into big economic gains for coffee growers. Birds consume coffee's

The Mexican guano bat is an insectivorous species that forms colonies of up to 20 million bats.

primary insect pest, a small beetle named the coffee berry borer. The coffee berry borer burrows into coffee berries and consumes them from the inside out. Across the world, the coffee berry borer may be responsible for more than U.S. $500 million in harvest losses annually. In Jamaica, birds consume this pest, resulting in big economic savings for coffee growers—up to $310 per hectare (2.5 acres) on one plantation.

In Costa Rica, the coffee berry borer is an important pest in coffee fields. Although it only arrived in Costa Rica around the turn of the twenty-first century, preliminary results suggest birds and possibly bats were already consuming significant quantities of the pest a decade later. When cages experimentally excluded birds and bats from foraging on a coffee plantation, coffee berry borer infestation doubled from 5 to 10 percent of the berries. The effect was not limited to the berry borer; insect abundances in general were much higher inside the cages than outside. Which bird and bat species are doing the bulk of the pest-control work is unclear but pivotal for farmers to know if they are to manage their land to ensure continued pest-control benefits. Fortunately, emerging techniques from molecular biology are allowing researchers across the globe to figure out what animals eat.

Coffee growers are not the only people who benefit from pest-controlling birds and bats. Attracting great tits to apple plantations in the Netherlands by putting out nest boxes has significantly increased apple yields. In the United States, insectivorous bats annually consume some U.S. $750,000 worth of cotton pests in just eight counties in Texas. Across the country, bat-mediated control of insect pests is valued at several billion dollars annually—a free service from nature that is now jeopardized by the spread of white-nose syndrome in bats.

Other agroforestry systems benefit from birds and bats as well. In the past few decades, oil palm cultivation has been rapidly expanding in the tropics, seriously endangering native biodiversity, particularly in Southeast Asia. Palm oil has many uses; in addition to its role in cooking, the oil is used in soaps, candles, and cosmetics, and increasingly as biodiesel. Globally, the land under oil palm cultivation has tripled since 1961. More than half of the oil palm expansion in

Malaysia and Indonesia has occurred at the expense of primary forests, devastating native plants and animals. The conversion of primary forests to oil palm plantations, which have been described as "biological deserts," results in the near disappearance of the vast majority of birds, mammals, butterflies, and other animals that evolved to live in the forests. The current rapid spread of oil palm plantations across the tropics therefore spells doom for much of the species-rich native forest biodiversity and generates great increases in emissions of global-warming greenhouse gases, but spells huge profits for oil palm barons and their paid propagandists.

Indeed, it would be quite appropriate to refer to oil palms as "blood palms" because expansion of their plantations undoubtedly jeopardizes the livelihoods of poor rural folks. Oil palm expansion in Indonesia, for example, has resulted in human rights violations through land-grabbing and deforestation. Even in government-run programs, local farmers are asked to give up 10 hectares (25 acres) of their land to oil palm companies and in return receive rights to only 2 hectares (5 acres) of the land under oil palm cultivation. The spread of oil palm plantations is also considered a threat to both cultural heritage and human health. The latter is due to the loss of land for cultivating food crops, thus putting local food security at risk.

Oil palm plantations occasionally suffer extensive defoliation from insect pest outbreaks, which can depress yields and hit the bottom lines of oil palm companies. To reduce the damage from insect pests, oil palm companies frequently use pesticides, risking harm to the environment and human health. Other approaches such as planting beneficial plants that attract predators and parasitoids of oil palm pests have also been attempted, but with limited success. When insectivorous birds were present on oil palm plantations, foliage damage was reduced by 28 percent in Borneo. Since defoliation and fruit yields are linked, this service by insectivorous birds could enhance the yield of fruit and oil. Some tropical rain forests might be spared the axe if yields from existing plantations could be increased. To accomplish that sustainably, oil palm companies could retain ground vegetation and epiphytic ferns on oil palm trees to attract insectivorous birds. In addition, preservation of nearby primary forests could help maintain the supply of insectivorous birds ("natural insect pest-busters") for the oil palm plantations. But oil palm planters have no such far-sighted ideas. Indeed, the greed of oil palm promoters is difficult to overestimate, and they run a worldwide campaign of disinformation to increase their markets.

Vultures, Rabies, and an Ancient Ritual

A recent dramatic event in India illustrates another connection between biodiversity loss and human welfare. Scavenging vultures have been serving as free janitors, very important ones, especially in developing countries with poor sanitation. The vultures feed on dead and decaying animal carcasses and, by so doing, assist in nutrient cycling and maintenance of the rich community of tiny animals, plants, fungi, and microbes that make soils productive. In addition, vultures attract other scavenging animals to carcasses, and together they all prevent the spread of disease by disposing of rotting carcasses quickly. Vultures locate carcasses by sight, often from miles away, and, in some species, by smell; they are thought to have the best sense of smell of any bird. Because vultures feed on animals that may have died of disease or poisoning, they obviously must have potent immune systems. The strong acid in their digestive system is believed to kill pathogens that they may pick up from the carcasses. Further, vultures urinate on their featherless legs; evaporation of the urine cools their bloodstream, and it has been hypothesized that the urine kills germs picked up while feeding on carcasses.

Since the 1990s, there has been a more than 90 percent decline in vulture populations in India, and similar population collapses have been reported elsewhere in Asia and in Africa. The veterinary use of the anti-inflammatory drug diclofenac appears to be behind these declines, at least in Asia. Cattle are considered sacred in India so when the animals are old they often are treated with diclofenac to help them cope with joint and muscle pain. Vultures ingest the drug when they feed on carcasses of livestock that were treated with it. Diclofenac residues in vultures result in renal failure and visceral gout (a buildup of uric acid in internal organs). Ornithologists have reported that breeding populations of the white-backed vulture have been obliterated from the Keoladeo National Park in India, possibly because of diclofenac-related effects. To reverse the decline of vultures, the Indian government banned diclofenac in 2006. Pharmaceutical companies in the Indian subcontinent are now promoting a new drug that is considered safe for vultures—meloxicam—as an alternative to diclofenac. However, we hope that any negative effects of this drug on biodiversity will be carefully evaluated before its use becomes widespread.

The decline of vultures appears to have resulted in population explosions of disease-carrying rats and feral dogs. India suffers one of the highest numbers of human deaths due to rabies. With feral dogs replacing vultures as carcass disposers, there have been concerns in India about the recent rise in rabies. The increase in cases of rabies and other diseases has amplified the repercussions of declines in vultures on human health. An outbreak of bubonic plague in India in 1994 is suspected to have been due to an increased population of rats infested with plague-carrying fleas. The rat population may have risen with the numbers of undisposed carcasses. In addition to fifty-four human deaths, the plague cost the Indian economy an estimated U.S. $2 billion for quarantine measures, human evacuations, and a loss of tourism dollars.

Pathogens such as those that cause rabies and canine distemper also can be transmitted by rats and feral dogs to a wide range of other potential host species such as mongooses and jackals, populations of which may increase as vultures decline in African savannas and as carcasses become more available to them. Cattle carcasses left to rot, it is feared, could also spread anthrax to livestock. Scavengers that replace vultures at carcasses thus may spread diseases to wildlife, livestock, and human populations.

The loss of vultures has other social and economic implications. For centuries, vultures have played an indispensable part in Parsi burial ceremonies. Parsis originated as Zoroastrians in Iran who migrated to India in the tenth century because of persecution by Muslims. The global population of Parsis is estimated at around 100,000. Parsis believe that if they burn a dead body, they will dishonor fire. And, if they bury the dead, they will pollute the earth. Therefore they use the Tower of Silence on which dead bodies are exposed to the skies and to the vultures, which eat the remains.

Most Parsis in India are now concentrated in Mumbai (formerly known as Bombay). They have been taking their dead to a parkland hill on the outskirts of the city. This place was originally secluded, but thanks to human population expansion, it is now surrounded by urban sprawl. It contains a number of Towers of Silence. Until the 1990s, hundreds of vultures could be seen perching on the towers, and they competently disposed of the bodies. Now the disappearance of vultures from Mumbai is causing a spiritual crisis among the Parsis. To counter the loss of vultures as the principal agents for disposal of the bodies, the Parsis are currently using solar panels to concentrate heat on the dead bodies and speed the process of decomposition. However, Parsis remain divided on whether their ancient burial ritual is outdated. Some argue that "sky burial" is archaic and should be replaced by cremation, while others have proposed that vulture aviaries should be erected over some of the towers.

Finally, the decimation of vulture populations is also impacting some traditional occupations. Vultures can clean a carcass quickly, leaving little more than bones, which are then gathered by bone collectors and sold to fertilizer, gelatin, and glue industries. But this is becoming almost impossible because of the lack of vultures; their decline thus has been jeopardizing the livelihoods of already impoverished bone collectors.

Defaunation

Three decades ago in tropical forests in Los Tuxtlas, eastern Mexico, and Guanacaste in Costa Rica, biologists noticed carpets of seedlings surrounding parent trees at their bases, similar to what Navjot Sodhi saw in Singapore. They too concluded that this unusual observation was connected to a lack of seed dispersers. The seeds had simply dropped beneath the parent plants, giving rise to the seedling carpets. Most of the seedlings died from fungal diseases and lack of sunlight and nutrients. Similar observations have subsequently been made in tropical forests in Asia and Africa, indicating that the phenomenon is widespread throughout the tropical world.

A term has been created to capture the effects of the loss of mammals and birds: "Anthropocene defaunation." This term describes many forests throughout the world that look magnificent, with impressive stands of trees, but that have lost most of their me-

Though it is still relatively abundant, the jaguar is considered "near threatened" by the International Union for the Conservation of Nature. It is the largest cat in the Western Hemisphere, and its remaining habitat tends to be submarginal and often fragmented. In addition, hunting and poaching are reducing the jaguar's prey, and it is persecuted as a danger to livestock.

dium-sized and large mammals and birds. In those forests, vital ecological processes such as pollination and seed predation and dispersal have been reduced or eliminated. Enter an empty forest and you will be met with silence. It is a weird feeling to walk for hours along a trail in such a forest without seeing a monkey or a squirrel and hearing only a few lonely bird calls. Where are these forests? They occur in the Amazon basin; Costa Rica; southern Mexico; Queensland, Australia; much of equatorial Africa, and Southeast

Asia—in other words, throughout the tropics wherever forests still exist.

A recent evaluation of defaunation throughout the world has provided a glimpse of the grim situation of populations and species of vertebrates. It is estimated that more than 50 percent of all wildlife has been lost in the last forty years; in regions such as South America the loss could be as high as 70 percent! The disappearance of large mammal predators has profound effects on their native communities. In the 1970s the large Guri dam was built 97 kilometers (60 miles) inland from the Orinoco Delta of Venezuela, creating several islands of various sizes, some of which lacked large predators such as jaguars. A natural experiment was created that provided critical insight about the role of predators in the ecosystem. There were many interesting, although sad, consequences of the isolation of the islands and the resultant loss of predators as well as other species. The disappearance of large predators led to what has been called an ecological meltdown. With no predators, the herbivores quickly multiplied, causing a dramatic decline in abundance of the seeds and saplings of canopy trees. The absence of jaguars led to profound changes in plant composition, including the local extinction of valued trees and other plants and an invasion of vines and other plants that overgrew the native plants. In addition, aberrant behavior was observed in animals like monkeys, which turned to cannibalism in overcrowded conditions. Losing the predators thus created an unanticipated cascade of ecological problems, illustrating the potential consequences of the extinction of predator populations and species in most natural ecosystems.

In tropical forests, the presence of megacarnivores such as jaguars limits the populations of small and medium-sized predators (mesopredators). Mesopredators become more numerous following the demise of top predators and, among other results, can cause havoc for nesting birds. For instance, a decline in numbers of harpy eagles in Central America has resulted in population increases of avian nest predators such as white-faced capuchin monkeys, which has exacerbated losses of eggs and nestlings of birds on Barro Colorado Island in Panama. Heavy predation on avian nests may lead to population declines, and even extinctions, of some forest bird species. Thus the disappearance of large predatory birds and mammals from an ecosystem has important downstream implications for biodiversity.

The collective ecological and economic impacts of the losses of countless animal species are huge, and defaunation is by no means confined to tropical forests. Yet the impacts and costs are often unanticipated and unappreciated. The suburbs of San Diego, California, have been a place for other "natural" experiments on the impact of predator loss on ecological communities. Development has destroyed much of the sage-scrub habitat and greatly fragmented the remnant areas, creating an archipelago of habitat islands. Suburbanization has decimated populations of coyotes, the top predator in the system (at least for the past century or so). That in turn, has released the most common mesopredators—skunks, raccoons, foxes, opossums, and feral house cats—meaning their populations, released from competition and predation from coyotes, have increased. As a result, the populations of scrub-breeding birds have suffered heavy predation and decline, and one could guess that mosquitoes and other insect pests benefited from the birds' bad luck. This previously unsuspected phenomenon is now known as mesopredator release; once the large predators are gone, the populations of smaller predators grow unchecked, often with dire environmental consequences.

In the eastern United States, the loss of large predators and strict hunting regulations have caused the population of white-tailed deer to soar to an estimated 50 million animals, with densities as high as 50 deer per square kilometer (0.4 square mile). The overabundance of deer, the conversion of forests to croplands and pasture, and the fragmentation of remaining forests to augment the abundance of deer for hunters have caused many problems. For example, the deer have depleted the understories of shrubs and trees, which is causing a sharp and worrisome decline of until-recently common songbirds that have lost the plants they normally nest in. Ticks carrying Lyme disease have multiplied because of the abundance of deer and small mammals on which they feed. This has resulted in significant human illness and deaths every year. Controlling the abundance of deer is the only way to reduce these negative environmental impacts. But it is politically impossible to reintroduce predators such as wolves because of the close proximity of the deer to human settlements.

What happens if a long-absent top predator is restored to an ecosystem? Of course, the answer will vary with the particular ecosystem and the particular predator and its behavior. A very few such situations have occurred—most ecosystems have undergone losses of species at all levels—but one dramatic example is what happened after wolves were reintroduced to Yellowstone National Park in Wyoming in the United States. More than sixty years after the gray wolf had been exterminated in the region, several individuals were reintroduced to the park in 1995–1996. By 2011 there were about one hundred wolves within the park, and the growing population had spread to areas far outside it. Within those fifteen years, significant changes in both the flora and fauna of the park had taken place. The wolves' principal prey, the elk herd, had been reduced by more than half, and their feeding patterns had markedly changed. Relieved of heavy browsing pressure that had prevented survival of seedlings, several plant species—willows, aspen, and cottonwood—began to recover. The return of willows fostered a rebound of beavers from one to twelve dens in the park. The beavers in turn have improved riparian habitats by creating dams and ponds and regulating stream flow, all to the great benefit of fish and waterfowl populations, as well as amphibians and reptiles. The increased tree growth has provided new habitat for songbirds, which have now returned to the park in greater numbers. The population of coyotes, however, has declined, probably reducing pressure on the populations of small mammals such as rabbits and ground squirrels, while providing more

food for mesopredators such as red foxes, ravens, and bald eagles. These unexpected cascading results of the Yellowstone wolf reintroduction experiment have vividly demonstrated the importance of predators in a complex ecosystem.

The Mexican wolf almost became extinct in the wild in the late 1970s. Fortunately, the Mexican and U.S. governments worked together to capture the last wild individuals in Mexico and started a captive breeding program. Three decades later, this wolf was released in Arizona. In 2013 they were reintroduced in Mexico, and in 2014 a litter was born in the wild for the first time in more than thirty-five years.

Local Extinctions and "Zombie" Populations

Most likely, both species extinctions and extinctions of local populations will become even more prevalent as human population growth, climate disruption, and the release of toxic substances accelerate. Because of continuing heavy deforestation, Southeast Asia will probably lose as much as 40 percent of its bird species by 2100. Indonesia, the country with the highest number of resident and endemic bird species in Southeast Asia (929 and 408 species, respectively), stands to lose the most bird species. Sodhi observed the disappearance of all primary forest in the Dieng Mountains in central Java over three years while he was doing research there in the early 2000s. The ramifications of the forest loss were felt in the village at the base of the mountains; villagers complained that there was no wood left to repair their huts and a lack of ecotourism because the site was no longer pretty. Villagers were also fearful of a black leopard that frequented the paddy fields, probably in search of food because it could not find it in the logged forest.

Unfortunately, the extinction predictions reported above may be optimistic as they do not consider the likely cumulative effects of other drivers of biodiversity change such as the increasing prevalence of huge fires in tropical forests (and other consequences of climate change), overharvesting, and invasive species. Similar, and perhaps better, projections need to be developed for other tropical areas.

One difficulty in making predictions is the persistence of "zombie" populations. It can take some time for mammals and birds to disappear following a habitat disturbance that ultimately dooms their populations to extinction. Possibly because the high environmental stability of tropical forests results in low adult mortality, tropical birds may have evolved comparatively low reproductive rates. Adult mortality of only 10 to 30 percent per year has been estimated for birds from various tropical locations. Such longevity suggests that some individuals of tropical bird species may survive for years following loss of their habitat as long as suitable food is available, unless that habitat loss itself reduces the adult lifespan. For example, about two thousand individuals of the endangered Puerto Rican Amazon (a parrot, not a river!), or Iguaca parrot, with an average life expectancy in captivity of 23 years, lived in the Caribbean National Forest during the early 1900s. However, even with protection, their numbers plunged to just twenty over the next sixty years. This indicates that it may take decades for a doomed population or species actually to disappear—it can persist as the "walking dead."

The Price We Pay

As should be clear by now, the widespread declines and extinctions of birds and mammals have disrupted ecosystem associations that have evolved over millions of years and at times have caused ecological meltdowns that significantly altered environments. The consequences of such extinctions are by no means restricted to tropical forests—they just are often especially dramatic in those settings. The functions of mammals and birds in benefiting the world's crops is worth billions of dollars every year. Tree crops (fruits and nuts) and vegetables need pollination assistance just as their jungle counterparts do. Even staple grains and other field crops benefit from the pest-control work done by birds and other animals. Yet those priceless services are too often overlooked or forgotten in an era of mechanical farming and chemical pest control.

So, in addition to the outcome of millions of years of evolution going down the drain, human health, cultures, and economies have suffered serious losses. Because such losses are usually local and unnoticed elsewhere, the cumulative impacts of worldwide extinctions are not fully comprehended. Not only are natural phenomena that have enormously enriched and sustained all of our lives disappearing, but with them have gone the chances for our descendants to enjoy similar benefits and pleasures. And this is just a sample of the repercussions of the negative impacts of our activities on the world's birds and mammals. Our future will be bleaker without the animals and plants that are disappearing and their fascinating and intricate interactions.

Not only are species and populations disappearing from Earth, but so are larger biological phenomena. The migration of wildebeest on the Serengeti Plains of East Africa is one of two remaining great mammal migrations on the planet, the other being that of Canadian caribou. The complex movement of more than 1.5 million individuals to find fresh grass after rains is now threatened by road development designed, among other things, to aid the exploitation of coltan. Here a group of some three thousand wildebeest plunge into the crocodile-infested Mara River in Tanzania in a spectacular, noisy, and risky crossing.

9 DRIVERS OF DEATH

VIEWED FROM SPACE, Earth seems to be an undisturbed planet, only sporadically shaken by a hurricane, or volcanic eruption, or some other large-scale natural phenomenon. Yet an astronaut can plainly see the planet's nighttime land areas illuminated like a Christmas display. Much closer, at the altitude of a small airplane, it is easier to see lands scarred by myriads of impacts derived from human activities. Our Earth is "a world of wounds," as the conservationist Aldo Leopold once described it. And worse yet, as he added, "it thinks itself whole."

Governments, nongovernmental organizations (NGOs), and individual citizens are making efforts, of course, to stem the tide of biodiversity loss. Many reserves have been put aside to protect other organisms, laws have been passed in an attempt to limit the exploitation of wild animal populations, and captive breeding programs have been set up in attempts to produce more individuals of severely endangered species to reintroduce into the remaining wilds. These programs have enjoyed some successes: wildlife populations supported by fees paid by hunters, California condors released into the Grand Canyon, Asiatic lions hanging on in the Gir Forest, and endangered regent honeyeaters managing to breed in Capertee, Queensland. But all of these successes are limited and likely only temporary in the face of increasing human population size and aggregate consumption, the basic drivers of environmental deterioration. Consequently, despite the efforts to protect them, the world's bird and mammal population and species diversity is disappearing. So let's consider the most basic drivers that must be altered or all the multitudinous other efforts will prove to be in vain.

The Human Dilemma

In this book we have only scratched the surface of the many ways in which the oversized human enterprise is demolishing the working parts of our planet's life-support systems. Some calculations indicate that, even with billions of people living in relative misery, another half an Earth would be required to maintain today's human population indefinitely. But those calculations do not fully account for the continuous and insidious erosion of the diversity of nonhuman populations and species—diversity that is fundamental to the health and welfare of *Homo sapiens*.

The impacts of human activity have passed from largely local and regional to dramatically global just in the past century or so. The human footprint is now visible in the most remote places of the planet, from the depths of the oceans to the highest mountains, on frozen tundra to the most impenetrable forests. Human activities are, without doubt, the cause of the sixth mass extinction now under way. But how did we end up here? The problems are complex, but the answer is simple, long known, and often denied: *too many humans consuming too many resources.*

Environmental scientists summarize it this way:

$$I = PAT,$$

where the *impact* (I) of the human enterprise on its life-support systems (including biodiversity) is the product of human *population size* (P) multiplied by

the average per capita consumption or *affluence* (A), multiplied by a factor incorporating the environmental impacts of the *technologies* (T) employed and how they are used to provide the consumption. For example, bicycles are a more environmentally friendly technology for transportation than automobiles, and car pools are more friendly than commuting as individuals; thus their I is lower than the I of a person driving to work alone every day.

Globally, the world seen through the lens of the I = PAT equation is a world where I is growing because all three factors are mostly growing. As the human enterprise expands, it displaces the natural world and its denizens. We are accelerating habitat fragmentation and loss, overexploiting living and nonliving resources,

The giant anteater is a toothless relative of the sloths, which, unsurprisingly, eats ants (and termites) by taking them into its mouth on its long, sticky tongue. It is widespread but has already become extinct in several countries, including Costa Rica. It cannot move quickly and thus is very vulnerable to hunting, anthropogenic fires, predation by dogs, road accidents, and habitat destruction.

The endangered Nilgiri tahr is a denizen of grassy hills in southern India. These goat-like animals were hunted for sport by European colonialists and today are threatened by poaching. A few thousand may still exist in the Western Ghats, and one group is reasonably well protected in Eravikulam National Park.

spreading destructive species into new areas, increasing the deposition of toxic chemicals from pole to pole, and playing risky games with the global climate.

The birds and mammals this book describes are suffering because we do all these things, which act synergistically to extirpate them. Thus forest fragmentation and climate disruption combined can have much more serious consequences than the sum of their separate impacts. Worse yet, the drivers of global impact are now correlated at many levels: human population growth causes increased demand for food and thus further clearing of forests for agriculture, which causes further changes in local or regional climates, both of which, in turn, decimate biodiversity. The loss of biodiversity then depresses agricultural production—and the race to the bottom accelerates. All of this does not paint a pretty picture of the future for the magnificent and fascinating birds and mammals of the world, to say nothing of our favorite mammal species, *Homo sapiens*.

It's all tied together. The more people there are, the more food society needs. The more food we eat, the larger and more intensified the agricultural system must be. The bigger the agricultural system (which is itself a major emitter of greenhouse gases from fossil fuel combustion, land-use practices, livestock production, and other factors), the bigger the impact on the climate. And birds and mammals? They are collateral damage, sentenced to death without malice aforethought, but disappearing all the same.

The sixth great extinction event in Earth's 4.6 billion-year-long history is likely to be accelerated by the same climate disruption that is increasingly challenging agriculture. The crises for biodiversity and agriculture are intertwined and, without exaggeration, deadly dangers to human civilization. Yet many people prefer to roll the dice, ignoring the abundant, carefully collected, and analyzed data. These are the deniers, and they will play their fiddles even as Rome's flames lick at their heels.

Population or Consumption?

Beyond the incurable deniers, misunderstanding of how demographic and environmental connections interact is common even among educated and reasonable people. For instance, many are convinced that overconsumption is a much larger contributor to environmental deterioration than overpopulation. This is roughly like being convinced that the length of a rectangle is a much larger contributor to its area than its width. The two factors are inseparable.

What typically happens during the development process in a preindustrial nation is that, with high death rates reduced by modern medicine, its population grows rapidly for a period, followed by a period of falling birth rates and slackening population growth and rising growth of per capita consumption. The fact that rapid growth of population and consumption do not occur completely in tandem is small consolation because the outcome is a large population with a gigantic amount of consumption and the further destruction of our life-support systems.

China is a recent example of this, as its previously skyrocketing population growth and its current skyrocketing growth in per capita consumption combine to make it a champion in wrecking the environment on local, regional, and now global levels, even perhaps outdoing the United States. While Chinese attitudes are changing, especially among a rising urban middle class, the nation has a long way to go. Asia's other huge country, India, is projected to add almost 400 million people by mid-twenty-first century and seems bent on following the superconsumption path using mid-twentieth-century technologies (e.g., coal-fired power plants). Similarly, more than 80 million additional Americans are projected to inhabit the already overconsuming United States by 2050; if they do, they will enormously augment the already massive U.S. attack on the planet's life-support systems. In sub-Saharan Africa, more than 1.1 billion people are projected to be added to the present 926 million by 2050, more than doubling the region's population. So many additional Africans too would greatly escalate the damage to the ecosystems they depend on. The Africa of our memories will probably be gone, and the remaining large animals, vestiges of the Pleistocene, will inhabit only zoos.

This entire situation is made worse by nonlinearities in the formula of population-consumption growth. Being clever, human beings use the easiest,

The chital, or axis deer, is not now globally threatened. It has a wide distribution in India and some neighboring nations; some of its native populations are large, and it has been introduced to many parts of the world, including Hawaii and Texas. Its range, however, is shrinking, and it plays an important ecological role as prey for such endangered predators as tigers, Asiatic lions, Indian wild dogs (or dholes), and leopards. As human overpopulation and climate disruption increase in the Indian subcontinent, the chital is likely to be hunted as a source of meat for hungry people.

most accessible resources first. This means that the richest farmland was plowed first and the richest mineral ores were mined first. Now each additional person must be fed from more marginal land and must use metals won from poorer ores. Thus, on average, each person added to the population disproportionately increases destruction of environmental systems.

The nonlinearities involved in resource extraction were dramatically underlined by the BP (British Petroleum) Deepwater Horizon blowout in the Gulf of Mexico in 2010. The first commercial oil well in the United States was drilled in Pennsylvania in 1859. It started at the ground surface and struck oil at 21 meters (69.5 feet). The Deepwater Horizon drill rig, 150 years later, started a well for BP in the Macondo concession in the Gulf of Mexico. Drilling began under water almost 1.6 kilometers (about 1 mile) deep and had penetrated almost 5 kilometers (3 miles) below the sea floor when the explosion occurred. The disproportionate impacts between the Pennsylvania and Gulf wells is just one sign of the diminishing returns that is one of the main harbingers of societal collapse. Such diminishing returns are now evident everywhere, affecting virtually all the resources civilization needs to persist, including once-vast populations of birds, mammals, fishes, and other living resources that humanity still attempts to harvest despite impending extinctions.

As the population grows, the efforts to keep people supplied with consumer goods will release more toxic compounds into the global environment. The toxification of Earth may be an even more dangerous trend than climate disruption, and may eventually contribute more to the extinction crisis.

Some people think that the population can be kept growing by improving the technology factor. There is, of course, much room for improvement in both efficiency and equity. For instance, largely abandoning personal vehicles for commuting and manipulating the economic system to reduce inequities (especially in food distribution) could greatly brighten the human prospect. But the history of claims that technological innovation will save us is instructive.

When *The Population Bomb* was published in 1968, the global population was 3.5 billion people, and many pronatalists claimed that technological innovation would allow society to give rich, fulfilling lives to 5 billion or more people. They would be fed by algae grown on sewage, whales herded in atolls, leaf protein, or food produced by nuclear agroindustrial complexes. None of that, of course, ever happened. The global population now exceeds 7 billion, and the number of hungry and malnourished people living today is roughly equivalent to the entire human population in the 1930s. We struggle to care adequately for those already on Earth, and any honest assessment quickly reveals that it will not be possible to feed, house, educate, and provide health care to billions more people.

The facts are uncomfortable; we are too many, and obtaining the resources most of us use in a typical day is too hard on the planet. Over the past century and a half, more than half of the environmental deterioration that has occurred can be traced to population growth, the remainder to growth in per capita consumption. For more than one hundred years, overconsumption has been largely powered by fossil fuels, and obtaining them now threatens local biodiversity even in remote places as oil rigs dot the north slope of Alaska and the depths of the Amazon basin, oil exploration threatens the megafauna of Murchison Falls National Park in Uganda, and coal mines despoil the once biodiversity-rich Tibetan plateau and ruin the agriculturally rich Hunter River valley of Australia. Because of the Amazonian oil wells, pipelines cross the Andes along routes cleared of tropical forests, and populations of endemic birds disappear. Closer to home, tar sands development is wrecking a large part of Alberta, Canada, and fracking for natural gas increasingly invades American suburbs, poisoning groundwater and the air with toxic substances.

Habitat Loss and Fragmentation

The International Union for Conservation of Nature (IUCN) estimates that more than 70 percent of all plants and animals at risk of extinction are being affected to a lesser or greater degree by habitat loss and fragmentation. This is one of the main drivers of species extinction. We have discussed the effects of habitat loss on a variety of mammals, such as orangutans, mountain gorillas, tigers, and Asian lions, and birds, such as condors, parrots, and eagles. When their habitat is fragmented, many species are unable to survive in the remnants submerged in human-dominated landscapes.

For example, only about 5 percent remains of the Mata Atlântica in eastern Brazil, a tropical forest ecosystem that is extremely rich in endemic species. The massive destruction suffered by the Mata Atlântica has caused severe population declines of many species of mammals and birds. One bird species is the Alagoas curassow, which likely became extinct in the wild in 1987. A captive population in Rio de Janeiro is the source of a reintroduction project. Many other species, including the maned sloth, the two endemic species of muriquis or woolly spider monkeys, the largest primates in South America, the black-faced lion tamarin, the recently described blond capuchins, the Brazilian merganser, the seven-colored tanager, and the red-browed Amazon, are on the brink of extinction due to habitat loss.

It would be immeasurably informative, from both a scientific and a management perspective, if we could hypothetically choose a representative country (ideally in the tropics because of the richness of their faunas and floras), magically allow it to fulfill its economic potential as standardly measured, and then document the consequent losses of bird biodiversity, all within a greatly accelerated time frame.

But this is not just a "thought experiment." The island of Singapore represents exactly such an ecological worst-case scenario—a "canary in a coal mine" that is giving us a dramatic warning, so far unheeded. Singapore has experienced exponential population growth from around 150 subsistence-economy villagers around 1819 to more than 5 million people today. Singapore has transformed itself from a third-world country of squatters and slums to a first-world metropolis of economic prosperity within the past several decades. Consequently, it is widely regarded by other developing countries as an ideal economic model.

Singapore's success, however, has come with a hefty price, one that was unfortunately paid heavily by its biodiversity. Since the British first established a presence in Singapore in 1819, more than 95 percent of the estimated 540 square kilometers (208 square miles) of original vegetation cover has been cleared, initially for the cultivation of short-term cash crops, and subsequently for urbanization and industrialization. As a consequence, Singapore has lost sixty-one of its original ninety-one forest bird species. It is suspected that many forest plants are not reproducing because

The southern muriqui is a species of monkey endemic to the Atlantic Forest, or Mata Atlântica, in southeastern Brazil. It is endangered because of habitat fragmentation and hunting.

of the loss of avian seed dispersal and pollination. Since Singapore is separated by less than 600 meters (2,000 feet) of sea from peninsular Malaysia, all these bird losses thankfully are local and not global species extinctions. Singapore is a microcosm of the tropical world, where forests are increasingly being fragmented by human activities such as urbanization, logging, and agriculture. And while Singapore is thriving by the usual economic metrics, it would not be in such good shape if depreciation of ecosystem services or self-sufficiency were factored into the analysis.

Numerous studies have shown that small remnant forests in deforested landscapes have poor conservation value for forest birds. A classic example of these "canaries" is the loss of bird species from the Barro Colorado Island (BCI) in Panama. BCI is a 1,562-hectare (3,800-acre) lowland forest that was isolated between 1911 and 1914 by the creation of Gatun Lake as a part of the Panama Canal. Birds on BCI have been regularly inventoried since 1929; 70 species (28 percent of the total) initially recorded have subsequently been extirpated, probably because of habitat deterioration and lack of recolonization by forest species. In other words, about nine forest bird species have been lost per decade. Additionally, a number of persisting species such as slate-colored grosbeaks have suffered severe population declines and may disappear in the near future. By comparison, the similar-sized (1,611-hectare, or 3,980-acre) La Selva Biological Station in Costa Rica has suffered comparatively fewer bird species losses (only 8, or 3 percent), probably because of its more recent isolation (15 years) and a corridor connection with a larger forest (44,000 hectares, or 100,000 acres), which doubtless permits augmentation of population sizes and re-establishment of locally extinct species through immigration.

Larger fragments, because of more heterogeneous habitats, are generally better refuges for forest birds than smaller ones. Small fragments can be penetrated by generalist predators such as crows, resulting in high predation on species not ordinarily exposed to them—a phenomenon called the negative edge effect. Changes in microclimate, a distinct climate regime in a very local area, can influence its habitability. Thus drier air circulating into a forest fragment from open surrounding areas is often deleterious to the forest's plants and animals and will affect a larger proportion of a small fragment than a big one. These effects seem to apply to forests in temperate regions as well as in tropical forests. Fragmentation of North American forests and the resulting penetration into forest fragments by nest predators such as raccoons, foxes, skunks, and crows, as well as brood parasitic cowbirds, may have been one of the main factors responsible for a large-scale decline of migratory birds that breed in the summer in North America and spend the winter in the American tropics.

Overharvesting and the Defaunation Syndrome

Overharvesting is one of the most pervasive drivers of animal and plant extinctions, and is responsible for nearly emptying many ecosystems of their larger birds and mammals. The immense appetite for animal products such as small birds for food and pets, elephant ivory, tiger bones, and monkey meat maintains an ever-growing industry of legal and illegal wildlife trade. Data from the NGO Traffic, an organization devoted to monitoring the wildlife trade, indicate that legal international trade in species of conservation interest included an annual average of 317,000 live birds, more than 2 million live reptiles, 2.5 million crocodile skins, 1.5 million lizard skins, 2.1 million snake skins, 73 tons of caviar, 1.1 million beaver skins, millions of coral pieces, and 20,000 mammalian hunting trophies, from 2005 to 2009. Traffic estimated that in the 1990s the value of legal wildlife products traded globally was around $160 billion; in 2009 the estimated value was more than $300 billion. Evidently, however, legal trade is only a part of the wildlife trade; illegal trade, much of it run by international criminal syndicates, is worth perhaps hundreds of billions more. As a prominent example, it is claimed that in the black markets of Southeast Asia, rhino horn is worth more than gold, cocaine, or heroin. Moreover, both the volume and the value of the wildlife trade, legal and illegal, evidently are increasing.

In addition to the trade of live animals and some of their products, the global scale of hunting for bush-

There are seven species of pangolins, like this one from Central Africa, and all are slow-moving animals that, if they cannot reach shelter, roll themselves into armored balls with erect scales to defend against predators. But the main predator is undeterred. Pangolins are hunted for bushmeat and for an international trade in skins and scales. The Chinese consider them a great delicacy and highly prize the scales for their putative medicinal action. Recently, local authorities seized a Chinese ship with 11 tons of illegally taken and killed Philippine pangolins.

meat is staggering, providing a harvest estimated to be millions of tons every year. Bushmeat is the chief protein source for people in many tropical countries in Africa and some parts of Asia, even though hunting bushmeat is mostly illegal. Thousands of gorillas, chimpanzees, bonobos, monkeys, wild pigs, deer, duikers, pangolins, and many other mammals are killed every year. Bushmeat hunters use various techniques from snares to guns to catch their prey. Snares, usually anchored cables or wire nooses, are widely used and extremely dangerous because they capture and cripple many species not intended for the bushmeat trade, including animals as large as elephants. In the once-splendid Nairobi National Park in the outskirts of Nairobi, Kenya, some 19,000 mammals are killed every year for the illegal bushmeat trade.

Although it is difficult to estimate the scope of the bushmeat trade, it is likely that millions of tons of wild animal meat are traded and consumed every year in Africa alone. An American journalist, Tim Butcher, in 2008 followed the route traveled by the famous explorer H. M. Stanley in the late nineteenth century. Butcher traveled by foot and motorcycle from Lake Tanganyika in Tanzania to the headwaters of the mighty Congo River, which flows 4,700 kilometers (3,000 miles) from East and Central Africa to the Atlantic Ocean. Butcher observed that for many miles the forests seemed devoid of life and concluded that this was caused by the relentless hunting of mammals and large birds for bushmeat. The magnitude of the African bushmeat trade has caused international alarm and has prompted some resolutions from

The gorgeous scarlet macaw is big: 91 centimeters (1 yard) long and more than 0.9 kilograms (2 pounds) in weight. It is also long-lived—more than half a century. Like other macaws, its beauty, highly valued for the cage-bird trade, threatens it. Scarlet macaws nest in holes in large trees, which tragically are sometimes cut down to gain access to nests. Reportedly, a baby scarlet macaw smuggled into the United States can be sold for as much as $4,000. Fortunately, the species has a very wide range from southern Mexico through Amazonia, and although habitat destruction has reduced some populations to remnants, it is still considered by the International Union for Conservation of Nature "of least concern."

international agencies. The famous primatologist Jane Goodall said almost two decades ago, "I firmly believe that unless we work together to change attitudes—from world leaders to the consumers of illegal bushmeat—there will be no viable population of great apes in the wild within 50 years."

The problem of overkilling is widespread and by no means confined to Africa. The numbers of animals killed are difficult to grasp. A few examples suffice to indicate the scale of the problem. In Brazil more than 24 million animals are estimated to be hunted every year. In Sulawesi, one of the larger islands in Indonesia, 90,000 animals are sold in a single market every year. And in Borneo 108 million are killed every year in Sabah, one of the smaller territories on the island. In Mongolia in 2004 hunters killed 3 million marmots and 200,000 gazelles, mostly illegally. The brutal killing is believed to have been even worse recently, leaving many of Mongolia's formerly idyllic grasslands literally emptied of their once-rich wildlife. In China the numbers of the chiru of the Tibetan plateau—the antelope symbol of the Beijing Olympics—have plummeted due to overharvesting for their wonderful fleece. An estimated 60 million birds and mammals are killed in Brazil every year. In the United States millions of animals such as muskrats, raccoons, opossums, and beavers are legally hunted for their fur, while other species considered pests, such as coyotes and prairie dogs, are exterminated with government subsidies.

Numbers alone are no protection against population or species extinction when the animals in question are economically valuable. For example, although there were billions of passenger pigeons, there was also a thriving market in the young birds (squabs) for food and in adults as living targets in shooting galleries. Whales and other sea mammals were once very abundant, but they were decimated when they were a source for lamp oil. Great whale populations were generally reduced to roughly 20 to 30 percent of their sizes before large-scale hunting; the northern right whale was nearly driven to extinction and remains endangered. Seals and their relatives were reduced even further a century ago by hunting for their fur, perhaps to as little as 3 to 5 percent of their previous abundance.

The akiapolaau is one of the few survivors of the once-splendid array of Hawaiian honeyeaters. Once there were more than fifty species of these fascinating birds; now less than half persist. It's a real thrill for a birder to see this endangered and scarce honeycreeper that fills the role played elsewhere by woodpeckers. It uses its straight lower beak to pry bark off, exposing holes containing insects, and its slender curved upper beak to hook them out.

Introduced Species

Successful introduced species, of both plants and animals, can have profound impacts on native animals, plants, and ecosystems. Indeed, introduced animals and plants may comprise up to 60 percent of the biodiversity in certain regions, such as the grasslands of California. Three decades ago it was estimated that 14 percent of all the plants throughout the world had been introduced to new locations by human action; things must be much worse now.

Sometimes, species that may be useful keystones and ecosystem engineers in their native habitats can do great harm elsewhere. An example we have met is the American beaver, native to Canada, the United States, and northern Mexico. But when introduced into Tierra del Fuego in South America, beavers caused tremendous damage by flooding extensive forests, changing the whole landscape, and threatening many indigenous species.

As we have seen, the introduction of novel predators is often responsible for the extinction of birds and mammals, especially on islands, such as Guadalupe Island in Mexico, Tahiti, Pitcairn, Hawaii, New Zealand, and Mauritius, which have been ravaged by introduced animals. A classic example of the havoc caused by invaders is that of the brown tree snake introduced in Guam in the western Pacific, discussed earlier. But the impacts of invasive species are in no way limited to islands. For example, it is certain that the introduction of European starlings to North America has affected many elements of the native avifauna, especially species that breed in tree cavities for which the invaders effectively compete. While a variety of other species, such as acorn woodpeckers, have been adversely affected by starling competition, it is still not clear whether any of their populations have been pushed toward extinction by the starlings.

Extinctions of Migrant Animals

Special problems are connected with efforts to protect migratory species. First, it is often important to preserve habitat for a species in two very different places—different nations or even different continents. Differing political attitudes on conservation between countries may pose greater difficulties than those entailed in trying to preserve most mammal species and challenges even more arduous than with nonmigratory birds. Mammals tend to be sedentary or to migrate in well-defined and well-understood patterns. Where their habitat needs to be protected is generally clear. For example, wildebeest, gazelles, and other mammals have for centuries migrated across the Serengeti Plains in Tanzania and the neighboring Masai Mara of Kenya, following the fresh forage provided by seasonal rains. These migrations are one of the great wildlife phenomena of Earth.

The migrations, in addition to being a treasure of biodiversity and an enormous and valuable tourist magnet, are now threatened by the planned construction of a highway bisecting the Serengeti Plains. The road would at least partially block the migration, give poachers easy access to game, and result in large amounts of road kill. Fortunately, a concerted effort was made to stop it, an amazing success since one of its purposes was to provide easy access to western Tanzania's deposits of coltan, an ore that can be refined into tantalum, a rare earth element that is a critical component of cell phones and computers. We hope that this is a permanent win since cell phones already are responsible for substantial losses of biodiversity as collisions with cell-phone towers annually slaughter millions of birds in North America alone. We jabber, and populations and species plunge toward extinction.

Migratory birds, however, while already facing hazards such as towers, tall buildings, wind turbines (a threat to insect-devouring bats as well), and pet and feral cats (which also kill billions each year), encounter additional problems. Migrating birds need habitat protection in both breeding and wintering ranges, and for many migrants there is an additional need for "refueling" stops during their journeys. Similarly, bats suffer from wind farms—it is estimated that an astounding 600,000 to 900,000 bats are killed every year by wind turbines in the United States. Some ameliorating steps are being undertaken to reduce mortality from wind turbines, such as improving turbine design or slightly raising the speed at which the blades start turning, so they present less of a hazard. This is critical, since the mobilization of solar power (wind is one form) will be essential for the survival of much biodiversity that is now threatened by climate disruption largely traceable to use of fossil fuels.

In many places, migratory birds face journeys made dangerous by gun-wielding poachers. A well-publicized example occurred in the winter of 2012, which was extremely cold in Eastern Europe. Some 50,000 migratory geese flying south to escape freezing, landed on Albanian flatlands where local market hunters exterminated them. Even tiny songbirds are not exempt from such slaughter. Every year billions of birds fly south from Europe across the Mediterranean to overwinter in the warmth of Africa. They are waylaid by hunters who shoot and trap them by the many millions. On Cyprus, during the migration, one songbird loses its life illegally every 4 seconds. By the end of the migration, millions will have died on that one small island.

Toxins as Influences

One of the most frightening problems facing birds and mammals (including us) is the pole-to-pole distribution of novel, slow-to-degrade, toxic chemicals synthesized by human beings—compounds lumped together as persistent organic pollutants (POPs) and endocrine-disrupting compounds (EDCs), which can have dire consequences in tiny doses. Some, such as DDT, which almost exterminated a number of bird species, are a long-recognized extinction threat. Yet now DDT is again the subject of a mendacious campaign by elements in the chemical industry that would like to start manufacturing it again, even though it is supremely dangerous to birds and other organisms and quickly becomes useless in most applications because pests rapidly evolve resistance to it. And recently, as we saw, a seemingly harmless anti-inflammatory drug used by veterinarians to treat cattle was a poison for vultures and produced a vast tragedy and widespread loss of an important ecosystem service.

Of course, not all the poisons with which *Homo sapiens* is assaulting wildlife are novel toxins the chemical industry synthesized. The element mercury, a by-product of coal combustion, is a perfect example. This nasty poison is now four times as common in the environment as it was a few centuries ago when coal was not being burned in huge quantities to fuel the industrial revolution. Mercury poisoning has been shown to alter bird songs, sometimes shortening and simplifying them—not surprising since it can also attack human brains and reduce verbal ability (among other serious neurological effects). The songs are crucial to bird mating behavior. Mercury, lead, and other heavy metals mobilized by human activities, especially coal burning, are now seen to be creating problems for avian populations worldwide—one more reason, on top of its major role in generating the greenhouse gas carbon dioxide, that humanity should move as rapidly as possible to cease mining and burning coal.

And in the offing, there is potential for catastrophe in the burgeoning production and global dispersion of nanoproducts, particles of diverse substances that can penetrate cells. The safety of these substances is essentially unexamined, but the results from a natural analogue, asbestos fibers, is not heartening.

So far no vast extinction tragedy from poisons has

afflicted mammals. The signs are not encouraging, however. Polar bears appear to be threatened by a family of nasty organochlorine compounds, PCBs (polychlorinated biphenyls), which tend to concentrate in colder areas like the Arctic and seem to deform the bears' gonads and weaken their immune systems. Many people have the impression that climate disruption is the worst environmental problem birds and mammals face, and indeed, its consequences may ultimately be catastrophic. But of other candidates for the biggest threat, the spread of toxic chemicals from pole to pole may be the dark horse in the race; Rachel Carson may have instigated environmentalism by illuminating exactly the right issue, referring in the title of her classic *Silent Spring* to the decimation of bird populations.

This is especially the case as recent research has shown that a long-held basic rule of toxicology, "the dose makes the poison," is often dead wrong. Yes, it's true that even table salt can kill you if you eat too much of it. If one plots the impacts of many toxic chemicals against the dose absorbed, there is a continuous increase of danger as the dose goes up. Such a dose-response curve is said to be monotonic.

EDCs may or may not do nasty things to you at high doses but can have different negative effects at extremely low doses. These low-dose adverse effects can increase the probabilities of altered sex determination and other developmental defects, behavioral problems, and development of cancers, among other effects. For these chemicals that behave like hormones, the response may first increase and then decrease as the dose size rises; the function may change sign from positive to negative or negative to positive. This happens for several reasons, most commonly because hormones and hormone-like chemicals alter gene expression at very low doses, but at higher doses the receptor-mediated responses are either saturated or shut down. This often produces U-shaped or upside-down U-shaped dose-response curves, which in the jargon of biology are non-monotonic dose-response curves (NMDRCs). Such nonlinearities are characteristic of many environmental situations, but as in toxicology, they are too frequently ignored.

EDCs have come into public-health prominence largely due to a 1996 book, *Our Stolen Future*. In the tradition of *Silent Spring*, the authors sounded the alarm, were attacked by the chemical industry, and were subsequently proven correct. Tyrone Hayes of the University of California, Berkeley, has suffered Carson-like abuse from agribusiness flacks for demonstrating nasty low-dose impacts of the near-ubiquitous herbicide Atrazine on amphibians in both the laboratory and in nature. Atrazine is manufactured by Syngenta, which had bankrolled a multi-million-dollar campaign hiring pundits to lie about Atrazine and besmirch Hayes's reputation. The campaign resembles that of the well-funded fossil-fuel industry assault on climate disruption as well as the tobacco industry's perpetual storm of lies about the safety of smoking. Many of the attacks on Carson focused on her gender; those on Hayes sometimes related to his skin color. He has been, as was Carson, scientifically vindicated and, like Carson, is one of our heroes.

Most of the symptoms of the potential problems of EDCs for biodiversity have been derived from studies of cold-blooded animals. Infant mosquito-fish subjected to small doses of 4-nonylphenol, an industrial chemical used in large quantities in various operations, produced adults that overall appeared to be females, although the normal 50:50 male-female ratio persisted in the sex organs. In the wild, a high frequency of alligators in a Florida lake polluted with EDCs had developmental abnormalities leading to sterility.

But there is anecdotal evidence that at least one mammal may have problems with proper sexual development as a consequence of exposure to EDCs, apparently influencing the interactions of hormones involved in sex determination. That mammal is *Homo sapiens*. For instance, in one Canadian village adjacent to a large complex of petrochemical plants the ratio of boys to girls at birth has recently declined from about one-half (normal) to one-third—although EDCs have not been demonstrated to play a role in that decline. In some parts of the Arctic, a notorious sink for EDC pollutants including PCBs, DDT, and flame-retardants, twice as many girls are being born as boys. And there apparently has been a general decline in the proportion of boys being born in Japan and among U.S. whites (but not African Americans)—due to more male fetal deaths.

Such findings raise questions about the potential effects of EDCs on birds and nonhuman mammals where data on sex ratios are very rarely available. Assigning blame to one or more chemical compounds or other factors in such cases is extremely difficult since many of them may interact to produce damage that is more than the sum of the individual damages. Direct effects and synergisms often defy identification, but they may already be affecting reproduction in natural populations of birds and mammals as Atrazine evidently is affecting frog populations. Toxins may eventually be responsible for a biological holocaust.

Climate Disruption

Finally, it should be obvious to you that climate disruption poses a dramatic threat to biodiversity in general as well as to people. This may be the most threatening problem ever faced by humanity. New predictions on the impact of climate disruptions warn that humanity may face what has been called the "end of civilization" before the twenty-second century. Climate change in the past was mostly gradual, and organisms could adjust either by migrating into more favorable climes as the situation changed or evolving the capability of thriving in new conditions. Climate disruption—extremely rapid and dramatic climate change, especially if generating more extreme weather events—makes such adaptation much more difficult or even impossible. Migratory birds that take their clues for movement from day length may arrive on breeding grounds too late to take full advantage of the seasonal flush of insects on which they depend for successful fledging of their young. The insects may take their cues from the time of snowmelt and emerge earlier and earlier as Earth warms. Not only that, extreme weather events can cause havoc for wildlife;

for example, in the winter of 2013–2014 insect population crashes associated with unusual bad weather in England devastated the food supplies of insectivorous birds and bats.

Similarly, migratory shorebirds depend on an abundance of organisms like horseshoe crabs and small fishes to allow them to "refuel" en route. If climate change causes such prey not to be at the right place at the right time in sufficient numbers, the birds will starve and be unable to complete their migratory journey. This is already happening in places as diverse as the Delaware shore, islands in the Arctic, and floodplains in Africa.

Migratory mammals in the Serengeti also may find themselves in the wrong place at the wrong time. Many changes in the distribution and abundance of animals and the timing of migrations, apparently in response to climate change, have already been documented, showing that a pernicious global process is under way. For instance, beetles that reproduce more rapidly in the warming climate are now killing great swaths of pine trees in the Rocky Mountains. They are producing vast areas of heavy fuel loads that will generate fires that could destroy much of the biodiversity in the affected regions, extirpating many bird and mammal populations.

And, of course, birds and mammals closely attuned to the environmental conditions of their habitats will find those habitats disappearing. Even if major components of the habitats are simply migrating, human disturbance may prevent many animals from following. Some plants (but perhaps not others) may easily disperse across urban barriers, and birds may fly over them, but some mammals may be unable to cross, thus shredding ecosystems. And birds and mammals that move up mountains to remain in suitable climate bands will eventually find themselves at the top and go extinct when they cannot climb higher. For example, it is thought that frog populations in the cloud forests of tropical mountains are already suffering extinctions from global warming.

All one really needs to remember is that many organisms, including most plants and birds and mammals, are tightly adapted physiologically to rather narrow sets of climatic conditions, and if those conditions change too rapidly or too dramatically, they are doomed. The consensus view, based on the 2009 Copenhagen Accord, has been that the safe limit of temperature increase is 2 degrees Celsius (3.6 degrees Fahrenheit), although at least one distinguished climate scientist considered that too high. Since that time it has become clear that the climate is changing more rapidly than expected and that various positive feedbacks are coming into play.

Those feedbacks are ones in which heating itself causes more heating. For instance, melting of arctic sea ice means less of the sun's energy is reflected back into space from the white ice; instead, more of it is absorbed by the much darker Arctic Ocean. The ocean is thus warmed further, melting more ice, causing more heating, and so on. Much the same can be said for large stocks of the powerful greenhouse gas methane trapped in arctic soils and oceans, which threatens to be released to the atmosphere by warming. Recent estimates are that the average global temperature could climb as much as 5 or 6 degrees Celsius (9 or 11 degrees Fahrenheit) by 2100. That would utterly transform the planet humanity is familiar with and would, in itself, be an utter disaster for biodiversity. Such a rise—too high and too fast—could destroy half or more of today's bird and mammal species and a much larger portion of their populations. Climate change alone could be sufficient to finish the sixth great extinction now under way.

Overall, climate disruption has multiple and complex impacts on birds and mammals. Indeed, one estimate is that half of all bird species are highly vulnerable to climate disruption, potentially causing human food shortages, habitat change and destruction, invasions of predators, competitors, and diseases. Both human societies and biodiversity are vulnerable to the direct effects of climate change: increased frequency and intensity of droughts, floods, violent storms, cold spells, heat waves, and other dangerous weather phenomena. The truth begs to be heard: we are the drivers of death.

Huge breeding colonies of the emperor penguin are established every year in the Antarctic. These colonies are suffering from the effects of global warming.

10 BEYOND MOURNING

PARAPHRASING THE WORDS of the poet A. E. Housman, the lessons we have to offer are not sweet; the truth is hard to hear. The conservation plight of birds and mammals is a story both tragic and shocking. Our generation is witnessing a disastrous decline among many of the world's best-known animals, including our closest relatives, the great apes. We are destroying some of the most fascinating and beautiful creatures that nature has ever designed, from singing whales and gorgeously colored birds to the weirdly human-like, sexually obsessed bonobos. The cultural and esthetic ecosystem services that these vanishing animals previously supplied will be denied to our descendants.

Even more staggering is the reality that birds and mammals, important as they are to humanity in so many ways, merely represent the tip of the iceberg of looming extinctions of virtually every other kind of organism on Earth. Indeed, some other groups are even more threatened than are our treasured birds and mammals. Thousands of plant species, for example, are at great risk, not least because of the disappearance of animals they relied on for their own reproduction. Perhaps the most startling news of all is that humanity itself is in deep trouble. All of us are affected; as populations and species go extinct, their disappearance further threatens the systems on which human life and civilization depends.

In addition to the innumerable roles diverse organisms play in supporting our lives, for many people the very existence of the only known other living beings in the universe is an esthetic and ethical service that cannot and should not be valued monetarily. There has been a long argument in philosophy about whether human attitudes toward biodiversity should be ecocentric (centered on the intrinsic values of nature) or anthropocentric (focused on the value of nature to humanity). We believe there is great potential in the ecocentric view—if people decided that biodiversity had value in and for itself, regardless of its instrumental value to *Homo sapiens*, the overall problem of preserving Earth's other organisms would be simplified.

At present, however, it has proven most useful for conservation biologists to impress on people that the future of the human enterprise rests to a substantial degree on preserving as much biodiversity as possible. The scientific community clearly understands the aggregate importance of biodiversity—just as one can comprehend the value of a police force without being able to specify the part each officer will play in suppressing a future crime wave, or the importance of rivets holding a wing on an airplane without knowing the specific purpose of each rivet. Recent efforts to assess the direct and indirect economic value of ecosystem services in general have been especially useful since we live in a global society in which, unfortunately, money has become the measure of almost everything. While the monetary values of biodiversity and ecosystem services are hard to estimate precisely, there is general agreement that they are gigantic.

Personally, we have been involved in many different activities to promote conservation of the plants and animals that make our lives interesting and satisfying. We have done basic scientific research to understand

The indri is a beautiful, distinctive lemur from eastern Madagascar. By now it will come as no surprise to you that hunting and deforestation threaten its existence on an impoverished island whose human population is projected to more than double by 2050.

how ecosystems are structured and function, as well as applied studies to develop new and better methods to assess the status of our fellow species and implement species conservation plans. Gerardo, for instance, has championed the creation of several biosphere reserves in Mexico that cover more than 2 million hectares (4.5 million acres) of tropical and temperate forests. Paul was largely responsible for establishing Stanford's Jasper Ridge Biological Preserve, which has hosted countless studies by students and faculty that have helped build the science of ecology. He and Anne have published extensively on global environmental issues, beginning with their 1968 book *The Population Bomb*, and have carried out diverse research on how to conserve threatened populations of butterflies, among other kinds of animals. The late Navjot Sodhi, who started to write this book with us, worked tirelessly to protect remnant tropical forests in Singapore and other regions in Southeast Asia.

Our work is not unique, but is similar to that of many thousands of individuals working for the conservation of biodiversity from local to global scales. Much more needs to be done to stem the losses of biodiversity, especially regarding global climate disruption, the ubiquitous spread of toxic chemicals, and the destruction of remaining natural ecosystems.

Of course, there is a limit to what can be achieved by personal, direct actions of any individual. Governments and other institutions must be involved. Many, if not most, environmental scientists are convinced that more than enough is known about the causes of the deterioration of human life-support systems to take appropriate corrective steps while still allowing us to lead comfortable, happy lives.

Rewilding and Countryside Biogeography

The Disney cinematographer Lois Crisler, after years of filming wolves in the Arctic, wrote, "Wilderness without animals is dead—dead scenery. Animals without wilderness are a closed book." Consider the idea of rewilding much of North America, a plan that restores both. If extensively implemented, rewilding would return "wolves, cougars, lynx, wolverines, grizzly and black bears, jaguars, sea otters, and other top carnivores" to their former homes in an ambitious process of restoring large, extensive ecosystems. We believe the approach of rewilding has considerable merit and would meet an ethical imperative not to exterminate other organisms, to preserve ecosystem services, and to reinvigorate North America's natural capital and productivity. Restoring healthy ecosystems also could help mitigate the impacts of climate change. We highlight rewilding as one hope, at whatever level it can be achieved, in our quest to save *Homo sapiens* and the birds, mammals, and innumerable other life-forms with which we share this planet. We applaud the effort, and we hope it succeeds.

A complementary, perhaps equally ambitious and difficult conservation measure comes from the discipline of countryside biogeography, created in large part by Gretchen Daily on the basis of her research in Costa Rica. The strategic focus of this new area of research is to make areas heavily impacted by humanity, especially agricultural areas, more hospitable to other organisms. Daily's Natural Capital Project aligns economic forces with conservation, supplementing and expanding the customary piecemeal approach to conservation. It greatly enhances the chances of innumerable species of plants, animals, and other organisms, known or obscure, to survive and continue supplying ecosystem services.

Unfortunately, although rewilding and countryside biogeography are laudable and far better than nothing, they have no influence on the principal drivers of the biodiversity crisis: continued growth of the human population and its consumption of resources. But they are not the only measures that have been taken to protect wildlife and natural ecosystems in the face of the human juggernaut.

Protected Areas

Setting aside areas for the conservation of scenic beauty and biodiversity has a long history; monarchs and princes used to maintain large tracts of natural habitat where native animals could flourish, often preserved for hunting and sometimes for their beauty. Most countries now have some at least partially protected areas or nature reserves, such as wildlife sanctuaries, national parks, biosphere reserves, indigenous peoples' reserves, and many other areas where the natural perpetuation of plants and animals is among the goals.

One of the most successful nature reserves in the world, the demilitarized zone (DMZ) between North and South Korea, was established essentially by accident. The undisturbed expanse of the DMZ has become an important overwintering refuge for the highly endangered red-crowned cranes, among many other birds, and numerous other animals

Closely linked to prairie dogs, which are its main prey, black-footed ferrets were presumed extinct in 1987, when a dog brought a dead specimen to its owner's home. Soon a small population was discovered in Wyoming, but then bubonic plague reduced that population to the brink of extinction. Finally, the last eighteen animals were captured and brought to a very successful captive breeding program.

persist there that have disappeared outside. It is a paradise of sorts. The lesson learned there is applicable to much of the world: remove humans and nature flourishes.

The vast array of different kinds of protected areas reflects the complexity of social, economic, and political conditions that now exist everywhere. The International Union for the Conservation of Nature (IUCN) advocates a goal of protecting at least 10 percent of Earth's land surface by 2020. Among nations, there is much variation, with a few countries such as Costa Rica having as much as 25 percent of their land protected, while others have very few or no reserves. In many regions, however, it is becoming extremely difficult to establish new reserves because most of the land is already under relatively intensive human use. Some countries such as Mexico, however, have ambitious, successful programs to increase the area of protected lands.

In the United States, the first national park was Yellowstone National Park. Set aside in 1872, primarily to preserve the last free-ranging buffalo herd and to protect the area's geological "wonders," especially geysers like Old Faithful, which erupts every hour, Yellowstone is genuinely a jewel of nature. In a single day a visitor can observe grizzly and black bears, wolves, elk, bison, pronghorn antelope, moose, deer, beavers, and many smaller mammals. More than a million visitors enjoy Yellowstone every year, and millions more visit the 384 other U.S. national parks and reserves. The United States pioneered national parks, but today some 1,200 parks and reserves exist in 100 countries around the world.

Visits to national parks were probably the first manifestation of ecotourism, an increasingly popular activity today that increases public appreciation of wildlife and nature and helps build support for conservation. Ecotourism can provide tourists (who often are key powerful people) with an introduction to a world generally suffering from a "nature deficit disorder." Indeed, there is no better way to appreciate biodiversity than to explore it in Africa where megafauna still roam, complete with easily observed large predators, a situation virtually unique today. Both conservation and ecotourism have become major activities in eastern and southern Africa, with substantial benefits both to the economies of these non-industrialized countries and to their abundant wildlife.

A heartening example in the area of ecotourism is the success of Botswana. A popular attraction is the swamp in the delta of the Okavango River, which receives floods annually from rains in the distant highlands of Angola. The water flows during the dry season from May to October into the world's largest inland delta, attracting regional and migratory birds and abundant game from the dry Kalahari Desert. The water eventually evaporates or is transpired from plants, leaving via the atmosphere rather than flowing into the sea.

In the Okavango Delta one can gain an appreciation of what much of the world was like before the loss of most of the Pleistocene megafauna. And for those with the means, it can be enjoyed in astonishing comfort. Not only is there the spectacle of large numbers of elephants and other big herbivores, but also substantial groups of that most attractive of antelopes, the sable. There one can have the pleasure of watching a group of oxpeckers, birds that make a living by eating parasites of large mammals, working over the hides of sables, searching systematically in rows for ticks.

Saving Endangered Targets

We have described only a sample of the plethora of species that have become extinct or are endangered because of human actions. One reason to be relatively optimistic, however, is that even critically endangered species can be saved from extinction by human actions. A number of methods are available to implement conservation of our vanishing companions, including breeding in zoos and specialized facilities, protecting their populations in the wild in sanctuaries or reserves, and management of surviving wild populations. Breeding animal species in captivity, including many bird and mammal species that have reached critically low numbers, is a conservation strategy that has worked quite well in some cases, but it is usually very time-consuming and costly and provides no guarantee of long-term success.

While captive breeding can help save and rebuild a population of a severely endangered species, it is somewhat pointless unless the population can be returned to its natural habitat. Examples of attempted and sometimes successful reintroductions have occurred throughout the world from captive-bred populations of endangered fishes to Asian cheetahs in Iran, Asian lions in India, California condors in the Grand Canyon, orangutans in Borneo, and red macaws in Mexico.

Some of these efforts have become controversial, however. The laboratory facility near Chengdu in China dedicated to the captive breeding and rearing of giant pandas has been quite successful, but it is not clear how much habitat can be protected to allow the long-term success of reintroductions. Indeed there is now the Chinese government's "Forest Tenure Reform" legislation, which in the worst case could destroy 15 percent of the panda's already inadequate habitat. Some biologists have claimed that, even though the pandas are great symbols of biodiversity, the effort is not worth the money, which would be better spent on conservation activities elsewhere. We disagree. First, the very existence of this effort may well help galvanize the Chinese government to protect panda habitat more effectively. Second, it is naïve to think that funds made available to save this charismatic species would be easily transferable. Finally (showing our ecocentric ethical concerns), we think Earth would be a poorer place without this wonderful animal—even if it only existed in zoos.

Nevertheless, it is clear that in most cases once a species has been secured in captivity or in a single park or reserve, its long-term conservation and instrumental value to people requires the transfer of some individuals to other places in order to increase the total numbers and establish additional populations. Establishing more than one population in more than a single location is an insurance mechanism that allows an increase in genetic variability and reduces the chance of problems that might wipe out the sole population. It is also a mechanism that could eventually allow restoration of ecosystem services to humanity in regions where they were impaired or lost when the local populations of that species were extirpated.

Large-scale movements and translocations of endangered (and other) species are becoming an essential conservation tool to cope with the ever-increasing negative human impacts on nature. These efforts are of three different kinds. Most commonly, individuals are transferred to introduce new populations or augment existing ones in the wild. Recently, for example, black-tailed prairie dogs have been reintroduced

New Zealand's takahe, a large flightless rail, represents one of a multitude of tame and often flightless birds wiped out when humans occupied the islands of Oceania a few thousand years ago. It too was thought to be extinct, but miraculously a small population was discovered in the Murchison Mountains of the South Island, while the takahes that once roamed the North Island are known only from their bones. It is thought to have declined first through climate change reducing its alpine habitat and then from the ecological devastation traceable first to the Maori and then the Europeans. But the species is slowly recovering, and takahes are even back in several places on North Island, transplanted to predator-cleared reserves such as Zealandia, where this one was photographed.

Symbolic of what we may be able to save is the heaviest arboreal relative of the common city pigeon, the giant Nukuhiva pigeon. This endangered wonder is surviving in populations of hundreds on two islands, Nukuhiva and Uahuka of the Marquesas in the South Pacific. It can have a wingspread of over three-quarters of a meter (2.5 feet) and weigh up to 1 kilogram (2.2 pounds). Suppression of hunting through agreements with local residents, especially on Uahuka (where it has been reintroduced), seems to be allowing the Nukuhiva pigeon to slowly recover.

The Tuamotu sandpiper has evolved away from the migratory habits of its relatives and away from typical sandpiper feeding behavior. Now declining in range and numbers, it is seriously endangered (probably about one thousand remain). This sandpiper persists on a few scattered low atolls in the Tuamotus that remain free of rats, foraging on dry sand and in shrubs and trees for insects and even nectar. Sea level rise will probably soon exterminate it.

from Mexico into Arizona, while black-footed ferrets and bison were reintroduced in the Janos Biosphere Reserve in northwestern Mexico from populations in the United States.

In Southeast Asia, moving orangutans has become the only hope for saving hundreds of individuals that become isolated in stands of a few trees when their surrounding forests are cleared for oil palm plantations. Capturing and transporting the isolated orangutans to places such as the Sepilok Orangutan Conservation Centre in Sabah has saved many of them. Once rehabilitated, they can be released again into the wild if suitable habitat is available and can be preserved in the face of accelerating deforestation.

Using Biodiversity Sustainably

Many species are endangered primarily because of direct pressure from human actions, such as hunting or collecting. In these situations, direct regulatory action may be required to save a population or species of particular economic value. Examples include valued fish species, some rare ornamental plants, and migrating ducks and geese. Sometimes measures such as captive breeding and transportation have been employed to save overexploited species.

A case in point is the desert bighorn sheep in Mexico, which has been under great pressure from hunting. In the 1990s, responding to political pressure from international conservation agencies, the Mexican government established a ban on hunting of the species. Before the bighorn sheep was protected, its population in Mexico fell to roughly 1,200 individuals. Understanding the complex dynamics among trophy hunting, activities of the local people, and the population of bighorn sheep was the first step in establishing a recovery program. That program included boosting the remaining populations in Baja California and Sonora, transferring some individuals to other states, and organizing a hunting protocol that would benefit local landowners and give them incentives to protect the bighorn sheep.

The original proposal met with fierce resistance, similar to that of ranchers in U.S. western states to the reintroduction of wolves in Yellowstone National Park and central Idaho in the mid-1990s. Working with the proud Seri Indians, the last nomadic tribe in North America, scientists in Mexico established an ambitious hunting and conservation program on Isla Tiburon, a large island in the Gulf of California. Bighorn sheep were introduced there in 1975, and by 1990 their population had grown to more than six hundred individuals. The right to hunt the first bighorn sheep legally on Tiburon under this new program was auctioned for U.S. $200,000. Bighorn sheep have been reintroduced in several other places, and the Mexican population now exceeds 13,000 animals. They clearly were saved from local extinction because of the incentives associated with the regulated hunting program.

Hunting of birds can also provide an important stimulus to conservation. Duck hunters in North America were concerned about how the loss of duck breeding habitat would affect their activities. In 1937 Ducks Unlimited was founded in Canada and now has branches in many countries. Since then it has conserved and restored more than 5.3 million hectares (13 million acres) of the most critical habitat for waterfowl and other wildlife. There is no question but that duck hunters in the past seventy-five years have helped reverse trends toward waterfowl population extinctions. Wetlands are among the most productive and important ecosystems, not only for breeding ducks and geese as well as myriads of other species, but also for an array of ecosystem services to humanity—most notably water purification.

Sometimes conservation programs can be difficult to maintain, however. For instance, there has been a heated debate in much of Africa over whether or not it is ethical to cull elephant herds that are growing in response to recent steps to stem the ivory trade. On the one hand, the giant beasts can be serious agricultural pests and at high densities can radically alter natural landscapes; on the other hand, animal rights activists and many other nature lovers are offended by the killing of these charismatic and intelligent animals.

Like many of today's ecoethical dilemmas, this one is not easy to resolve. There are ways to attain needed population reductions other than culling, including relocation and contraception. But suitable areas in which to introduce elephants are growing scarce, and using contraceptives is difficult except in small parks and is more complicated and expensive than shooting. Animal rights groups are, in our view, appropriately concerned about cruelty to elephants, and the plight of young elephants orphaned when their mothers are killed is especially heart-rending. But overpopulation of elephants can lead both to problems of sustainability for them and to collisions with another overpopulated species that has the capability of destroying them.

The Zimbabwean Campfire (Communal Areas Management Programme for Indigenous Resources) program, partially funded by USAID (United States Agency for International Development), was the center of a controversy in the 1990s over elephant hunting. The U.S. funding was to increase the ability of local people to manage natural resources. In this case, elephant herds outside parks and reserves were capable of ruining a family's livelihood by destroying its garden plot in a few minutes. Naturally, poor people shot raiding elephants. Worse yet, rogue elephants killed hundreds of human beings annually. Killing in defense of gardens and people was accelerating the decline in elephant herds, then due primarily to poaching.

The Campfire program sold some 100 to 150 licenses per year, for $12,000 to $15,000, to sport hunters to kill elephants. The returns were donated to rural district councils, which determined how the money was spent. Elephant herds grew dramatically in the hunting areas because poaching was suppressed by the elephants' new "owners," the local people, who got more money and suffered less damage. It was a win-win situation. But the Humane Society of the United States objected, saying that elephants should never be hunted, and animal rights groups successfully lobbied to get funding stopped.

More recently, despite the shocks of a cessation of international funding and the political-economic

implosion of the Zimbabwean state, the conservation benefits of Campfire have remained remarkably robust, although their present status is in doubt. The situation underlines the need to keep the ecoethics of the "big picture" always in mind, such as the impact of political disputes on biodiversity, and to pay attention to factors such as the disparate capabilities of different nations to protect the same species, when determining where to allocate conservation funds.

The Campfire controversy highlights the ethical conflict between those who believe the key conservation issue is maintaining healthy wildlife populations and those concerned primarily about the rights of individual animals or who decry the utilization or commodification of nature or, as it's sometimes called, wise use or multiple use.

African wild dogs are pursuit predators that hunt in large packs, following their prey for long distances, often at high speeds. Wild dogs are highly endangered, with numbers once in the hundreds of thousands now reduced to a few thousand.

Yet in the face of disagreements over values, we still have to make choices. Much as we hate to see elephants hunted by people who simply get a feeling of accomplishment from killing these magnificent animals, on the whole we side with the Campfire program. It seems better to give local people a beneficial stake in maintaining the herds instead of permitting their extermination than it does to avoid the "unethical" killing of nonhuman animals by rich hunting enthusiasts, especially where entire elephant populations may be doomed in the absence of hunting revenues. It would also show indigenous people that not all conservation programs operate against their perceived interests.

In addition, we think it is ethical to consider the non-charismatic animals and plants that, as we have seen in the field, can be laid waste by elephant overpopulation even while some other organisms can be dependent on normal elephant activities. Overall, regulated hunting seems to have a secure place in conservation, and hopefully will be increasingly supplemented or replaced by photographic safaris.

Nongovernmental Organizations, Legislation, and Public Opinion

In the United States, environmental nongovernmental organizations (NGOs) have been trailblazers in protecting biodiversity. Early efforts were based on the vast overexploitation of a resource few today would recognize as such: the feathers of birds. In 1886 the conservationist Frank Chapman famously took two hikes in Manhattan to the center of the women's fashion district and recorded forty species of native birds—all identified by the feathers adorning ladies' hats. The feathered-hat craze around the turn of the twentieth century was enriching plume (feather) hunters, who raided shorebird and gull rookeries along the U.S. East Coast. Professional hunters nearly exterminated snowy and great egrets, which were especially prized for their soft, long, breeding plumage. The avian holocaust, killing millions of birds annually, gradually generated outrage in the American public. In response, societies were organized to protect the birds and laws were passed to regulate hunters and the plume trade. Those societies took the name of that great pioneer painter of birds, John James Audubon. The consolidated Audubon Society and its local branches are still a force in North American conservation.

The plume-hunting episode marked the beginning of conservation in the eastern United States. In the West, also late in the nineteenth century, the wandering naturalist, philosopher, and author John Muir was creating a national consciousness of the beauty of the Sierra Nevada and similar natural wonders. He was largely responsible for the establishment of Yosemite National Park in 1890 and was instrumental in subsequently setting up several other western parks. For that reason he is often given the unofficial title of "father" of the U.S. national park system (even though Yellowstone National Park was established first in 1872). Perhaps equally important was his role in the founding of the Sierra Club in 1892 and becoming its first president. Growing beyond its early existence as an outings club, by the late twentieth century the Sierra Club had become a powerful force for conservation in the United States.

The beginning of the twentieth century was the time of America's first conservation-oriented president, Theodore ("Teddy") Roosevelt. He was a big-game hunter who had been appalled at the overgrazing and other environmental destruction he had observed in the West. When he became president in 1901, Roosevelt wielded his presidential power to protect wildlife and public lands. He brought the U.S. Forest Service into being and established 51 Federal Bird Reservations, 4 National Game Preserves, 150 National Forests, and 5 National Parks, and used the 1906 American Antiquities Act to create 18 National Monuments. Overall, Roosevelt protected more than 80 million hectares (200 million acres) of public lands and established a reputation for the Republican Party as an agent of environmental protection that was later extended by Richard Nixon, but was then slowly dismantled by Ronald Reagan and subsequent Republican leaders in Congress.

Increasing problems with pollution in the mid-twentieth century gave rise to popular resistance and the establishment of many more environmental organizations, which have since gained strength and continued to provide leadership on conservation issues. It is horrifying to contemplate what would be left of Earth's biological riches were it not for the World Wide Fund for Nature (WWF), The Nature Conservancy, and Greenpeace, as well as the older Sierra Club and the Audubon Society, and other groups that have long struggled against the tide of extinction, first in the United States and Europe, and now throughout the world. And new groups, including many local or regional ones in all nations, are joining the battle all the time.

Environmental NGOs, along with conservation biologists, zoos, and government entities such as fish and wildlife agencies and national parks, have been doing the critical job of slowing the rate of extinctions wherever possible and thus conserving some of humanity's most important resources, one that could provide the basic material for restoring our natural capital to its previous abundance. NGOs have also been instrumental in establishing international agreements to limit the biodiversity holocaust, such as CITES, the 1975 Convention on International Trade in Endangered Species of Wild Fauna and Flora. CITES was intended to prevent international trade in endangered species from wiping out species in the wild.

The MAHB

If the world economic system is to be revised, more attention needs to be directed toward human behavior, both individual and corporate. In response to this need, one emerging institution may help bridge the gap between individual action and institutional change: a movement called the MAHB, pronounced "mob" so that participants can be known as "mahbsters" (http://mahb.stanford.edu/). The acronym stands for Millennium Alliance for Humanity and the Biosphere. Among its goals is to create a new paradigm, foresight intelligence, a field of study focused on systematically examining likely or possible futures and then, critically, determining and helping implement the behavioral and institutional changes necessary for humanity to ensure a sustainable and equitable future for all.

To do this, MAHB hopes to foster a dialogue and collaboration between biophysical and social scientists and scholars in the humanities, and the rest of civil society, to address the critical issues facing society, including the crucial loss of biodiversity, and bring action to bear on all of them. The issues include disastrous trends such as the continued growth of the human population, the toxification of the planet, and overconsumption by the rich and the power-wielding plutocrats who developed the global system and maintain it for their personal profit with not the slightest concern for our fellow living beings. All of these, of course, also contribute to the loss of biodiversity.

The MAHB is attempting to develop a social movement committed to reversing the environmental and social trends threatening humanity and, above all, to emphasize the roles of the population growth and still-increasing consumption among the already well-off as drivers of environmental destruction. It also focuses on the sad effects of economic, racial, and gender inequity. Existing civil society groups and organizations are being enlisted to join forces, strengthen and build networks, share information, and act together for greater impact without losing their individual identities and missions. Among the MAHB's ambitious strategic goals is to bridge the gap between scientific knowledge and policy action and try to alter public perceptions of the issues and of nature, generate foresight intelligence to make society "future smart," and understand why societal responses to desperate conservation needs are so severely limited.

A Losing Battle?

But, important as the conservation NGOs and government entities working to protect biodiversity within nations are, current conservation efforts are only making the hemorrhage of biodiversity somewhat less severe than it otherwise would be. Overall, the efforts to protect Earth's treasure of biodiversity are failing. All indications are that conservation efforts are not even preventing the rate of loss from increasing. Every day multitudes of populations disappear, and every few hours several species are likely lost.

Worse yet, climate disruption and the continuing accumulation of toxic chemicals that now pollute Earth everywhere seem certain to accelerate extinction rates. Climate disruption (including ocean acidification) is a massive direct threat because many organisms will not be able to evolve or migrate successfully as their environments quickly become inhospitable. It is also a substantial indirect threat because of humanity's need to transform its energy infrastructure, revise its distribution of resources (especially water), and shift agricultural activities geographically in response to changing conditions over coming decades to centuries.

All that means more destruction and fragmentation of Earth's scarce remaining relatively pristine habitats, whether it involves covering deserts with solar installations, ruining near-shore oceanic areas with the effluents from desalinization plants, building new canals and dams, or cutting down forests to grow crops for biofuel. This means that increased conservation efforts are not only essential, but that they will become even more so as people understand more clearly the importance of preserving natural habitats for sequestering carbon. In the face of these new demands on nature, the present level of conservation efforts may well be insufficient to avoid leaving a biologically impoverished world to our descendants.

What should be done? Rather than simply mourning the loss of Earth's biological riches, conservationists and scientists should be stressing their critical life-support importance to the general public, advocating for the preservation of biodiversity with governments and industry, and encouraging people to experience the esthetic and ethical services they provide. Of course, it is important that the piecemeal and incremental steps, like those we've described here, continue to be pursued in order to retard the ongoing destruction of our natural heritage as much as possible. Setting aside, restoring, and protecting more areas of natural habitat can establish reservoirs of populations and species that may prove invaluable sources of genes and organisms for further restoration. Relatively small changes in agricultural areas, such as encouraging the growth of a few trees or brushy field borders, can greatly enhance an area's capacity to support populations of native species. Even replacing lawns with natural vegetation in desert communities like Phoenix or Los Angeles can be helpful. That could also save precious water and, perhaps most important, help educate children about the beauty and interest of their local flora and fauna. The numerous conservation organizations and government agencies that pursue all these activities and policies should be strongly supported. Even though it is essentially a rear-guard action, it helps buy humanity time to address the root causes of our dilemma.

Beyond Conservation

What should be done to save the birds, mammals, the rest of nature, and ourselves? The only real hope is taking direct action to reduce the key drivers of extinctions and environmental degradation: overpopulation and overconsumption. The only effective measure is a rescaling of the human enterprise. As long as the human reproductive extravaganza continues and our material consumption addiction expands unchecked, birds, mammals (including humanity), and most life-forms will continue to die by a thousand cuts. Long-term hope lies only in shrinkage of Earth's human population and reduction of per capita material consumption among the already-rich. As long as people—especially business leaders, politicians, and economists—keep society hooked on perpetual growth, humanity will continue to cruise toward catastrophe.

Is there a way to avoid that lethal outcome? The answer is yes, but it requires much more than the biodiversity bandage approach tried so far. We need a global commitment to a brighter future, in which the

The aye-aye is an endangered treetop-dwelling, nocturnal lemur, a primate relative of chimps and human beings. Like all other lemurs, it is endemic to Madagascar. Sadly, the aye-aye was considered a symbol of bad luck by some local groups and was often killed on sight. That, along with habitat destruction, has been the principal reason for its decline.

fundamental disease is tackled as well as the symptoms. That means intensive action within nations to deal with both overpopulation and overconsumption. It would, first of all, involve steps to reduce human family sizes so that population growth ends and a gradual decline begins as soon as possible. That might be accomplished if the political will could be generated globally to give full rights, education, and opportunities to women, and provide all sexually active human beings with effective birth control. The degree to which those steps alone would reduce fertility rates is controversial, but they are a likely win-win for all societies. Obviously, especially with the growing endarkenment (the decline of Enlightenment values), especially in the United States, there are huge cultural and institutional barriers to establishing such policies. After all, there is not a single nation where women are truly treated as equal to men.

Population control, if it is to be accomplished

humanely, will be a slow process. But we know from experience that consumption patterns can be changed virtually overnight. That was shown clearly in the mobilizations and demobilizations connected with World War II. A stunning shift from civilian to military production occurred over some four to five years, and during those years Americans endured a great reduction of available durable goods and accepted rationing of gasoline, sugar, and meat. Other countries went through similar transitions, showing they too could make rapid changes in consumption patterns when the need was clear. Given appropriate incentives, economies clearly can be changed profoundly and very quickly. Obviously, to initiate a similar (but more gradual) change now, before the public is fully aware of our predicament, with all the attendant social disruption, will take political courage. One of the goals of this book, of course, is to ease that task by making the predicament more apparent to more people.

One of the saddest aspects of the situation is that, even if consumption patterns were dramatically changed, continued population growth alone could eventually bring down civilization. But reducing birth rates humanely in regions where they still are high to "below replacement" levels and eventually beginning to reduce the population size would take a decade or two to produce a significant change in today's population trajectory, and more than a century to reduce the human population size to one that could be sustained in a relatively long term. But that is no argument for delaying the effort. The population driver should not be ignored simply because limiting overconsumption can, at least in theory, be achieved much more quickly. The difficulty and lag-time involved in changing demographic trajectories means that the problem should be seriously addressed sooner rather than later.

Of course, ending population growth inevitably will lead to changes in age structure, with an increase in the proportion of older people, but that is no excuse for bemoaning falling fertility rates, as is common in some European government circles. Reduction of population size in overconsuming nations is a very positive trend from the perspective of both society and the birds and nonhuman mammals we care about. Sensible planning can deal with the problems of population aging. So we need to gradually and humanely reduce the size of the human population. By how much? Probably to where we were in 1900 or 1920, in the vicinity of 1 to 2 billion people, but society should have many generations of shrinkage during which to make that decision.

Much more quickly we also need to get a handle on our obscene consumptive behavior. How can we get off this insane, self-destructive treadmill? Each of us should, of course, make a difference by being less environmentally destructive, consuming and wasting less, having no more than two children (preferably one per couple), seeing that every sexually active individual has access to modern contraception and backup abortion, supporting women's rights everywhere (which not only is simple justice but helps lower birth rates and provides major benefits to society), promoting more equitable distribution of wealth within and between nations, and appreciating and supporting biodiversity with action or money.

Toward a Better Future

We've come to the end of our trip through the fates and futures of our most familiar living companions, birds and mammals. In this book we have outlined a grim account of the state of affairs on our planet. Our purpose, though, is not to leave you in the depths of despair. There is still enough biodiversity, and a bit of time, to avoid the worst of our possible fates. They are still out there, the birds, mammals, and most of the other flora and fauna, and we are still here. Take action. Go to your backyard, a nearby woods, or the Serengeti, and see some of those wonderful creatures. Join the Millennium Alliance for Humanity and the Biosphere and collaborate with others who are attempting to rally civil society to save biodiversity and civilization.

Does civilization have the will and wisdom to change our ways and permit adequate living room for our only known companions in the universe—companions vital to our physical, esthetic, and ethical interests? Today the road we are on suggests the answer will be a thoughtless "no." But a better kind of future can be chosen, and the choice is ours.

Saving all the species that have accompanied us in our journey in the huge, cold, and little-comprehended universe depends, exclusively, on us. Paradoxically, our future depends, inexorably, on the fate of these marvelous creatures. There are many reasons to save them. But as the famous French naturalist Jean Dorst once said, "Nature . . . will only be saved if man shows it some love simply because it's beautiful. This is also part of the human soul." Let's cling to the possibility that humanity will reverse course, and birds, mammals, and all the other wonderful creatures and vegetation will remain varied and abundant and able to support our lives, feed our esthetic senses, and boost our spirits. Above all, let's all work to make that possibility a reality.

Appendix: Common and Scientific Names of Plants and Animals Mentioned in the Book

Adélie penguin (*Pygoscelis adeliae*)
African elephant (*Loxodonta africana*)
African forest elephant (*Loxodonta cyclotis*)
Agouti (*Dasyprocta* spp.)
Akohekohe (*Palmeria dolei*)
Alagoas curassow (*Mitu mitu*)
American bison (*Bison bison*)
Arctic tern (*Sterna paradisaea*)
Arunachal macaque (*Macaca munzala*)
Asiatic cheetah (*Acinonyx jubatus venaticus*)
Asiatic lion (*Panthera leo persica*)
Atitlan grebe (*Podilymbus gigas*)
Australian night parrot (*Pezoporus occidentalis*)
Bachman's warbler (*Vermivora bachmanii*)
Bald eagle (*Haliaeetus leucocephalus*)
Bald parrot (*Pyrilia aurantiocephala*)
Bali tiger (*Panthera tigris balica*)
Barashinga (*Rucervus duvaucelii*)
Barbary lion (*Panthera leo leo*)
Barn swallow (*Hirundo rustica*)
Bird of paradise (*Parotia berlepschi*)
Black-backed jackal (*Canis mesomelas*)
Black crowned crane (*Balearica pavonina*)
Black-faced lion tamarin (*Leontopithecus caissara*)
Black-footed ferret (*Mustela nigripes*)
Black rhino (*Diceros bicornis*)
Black-tailed prairie dog (*Cynomys ludovicianus*)
Blond capuchin (*Sapajus flavius*)
Blue crane (*Anthropoides paradiseus*)
Blue whale (*Balaenoptera musculus*)
Blue-winged macaw (*Primolius maracana*)
Bonobo (*Pan paniscus*)
Bornean orangutan (*Pongo pigmaeus*)
Bowhead whale (*Balaena mysticetus*)
Brazilian merganser (*Mergus octosetaceus*)
Brazilian tapir (*Tapirus terrestris*)
Broad-faced potoroo (*Potorous platyops*)
Brown bear (*Ursus arctos*)
Brown tree snake (*Boiga irregularis*)
Buffalo (*Bison bison*)
Bugun liocichla (*Liocichla bugunorum*)
Burrowing owl (*Athene cunicularia*)
California condor (*Gymnogyps californianus*)
Cape buffalo (*Syncerus caffer*)
Cape lion (*Panthera leo melanochaitus*)
Caquetá titi (*Callicebus caquetensis*)
Caraiba (*Tabebuia caraiba*)
Caribbean monk seal (*Neomonachus tropicalis*)
Carolina parakeet (*Conuropsis carolinensis*)
Caroline Islands reed-warbler (*Acrocephalus syrinx*)
Caroline Islands swiftlet (*Aerodramus inquietus*)
Caspian tiger (*Panthera tigris virgata*)
Cats' claws (*Uncaria* spp.)
Cazahuate tree (*Ipomoea arborescens*)
Central hare-wallaby (*Lagorchestes asomatus*)
Cheetah (*Acinonyx jubatus*)
Chimney swift (*Chaetura pelagica*)
Chimpanzee (*Pan* spp.)
Chimpanzee, common (*Pan troglodytes*)
Chinese river dolphin or baiji (*Lipotes vexillifer*)
Chinstrap penguin (*Pygoscelis antarctica*)
Chital deer (*Axis axis*)
Christmas Island pipistrelle (*Pipistrellus murrayi*)
Christmas Island rat (*Rattus nativitatis*)
Christmas shearwater (*Puffinus nativitatis*)
Clouded leopard (*Neofelis nebulosa*)
Cocktail ant (*Crematogaster sjostedti*)
Coffee berry borer (*Hypothenemus hampei*)
Collared peccary (*Pecari tajacu*)
Colorful babbler (*Bugun liocichla*)
Common (black) rat (*Rattus rattus*)
Common carp (*Cyprinus carpio*)
Common nighthawk (*Chordeiles minor*)
Coral tree (*Erythrina* spp.)
Cotinga (*Cotinga* spp.)
Coyote (*Canis latrans*)
Crab-eating macaque (*Macaca fascicularis*)
Crescent nail-tail wallaby (*Onychogalea lunata*)
Cross River gorilla (*Gorilla gorilla diehli*)
Crowned woodnymph hummingbird (*Thalurania colombica*)
Cryptic forest falcon (*Micrastur mintoni*)
Cuban macaw (*Ara tricolor*)
Curassow (*Crax* spp.)
Curlew (*Numenius* spp.)
Darwin's fox or warrah (*Dusicyon australis*)
Deer mice (*Peromyscus* spp.)
Desert bighorn sheep (*Ovis canadensis nelsoni*)
Desert rat-kangaroo (*Caloprymnus campestris*)
Dodo (*Raphus cucullatus*)
Duck-billed platypus (*Ornithorhynchus anatinus*)
Dunlin (*Calidris alpine*)
Eastern gorilla (*Gorilla beringei*)
Eastern hare wallaby (*Lagorchestes leporides*)

Eastern lowland gorilla (*Gorilla beringei graueri*)
Elephant bird (*Aepyornis* spp.)
Elk (*Cervus canadensis*)
Emu (*Dromaius novaehollandiae*)
English sparrow (*Passer domesticus*)
Eskimo curlew (*Numenius borealis*)
Eungella honeyeater (*Lichenostomus hindwoodi*)
Eurasian beaver (*Castor fiber*)
Eurasian bee-eater (*Merops apiaster*)
European starling (*Sturnus vulgaris*)
Falkland island fox (*Dusicyon australis*)
Feral house cat (*Felis catus*)
Ferruginous hawk (*Buteo regalis*)
Flightless cormorant (*Phalacrocorax harrisi*)
Flying lemur (*Cynocephalus volans*)
Forest elephant (*Loxodonta cyclotis*)
Frigate-bird (*Fregata* spp.)
Fulmar (*Fulmarus* spp.)
Galapagos penguin (*Spheniscus mendiculus*)
Gentoo penguin (*Pygoscelis papua*)
Giant panda (*Ailuropoda melanoleuca*)
Gibbon (Family Hylobatidae)
Golden capuchin (*Cebus xanthosternos*)
Golden eagle (*Aquila chrysaetos*)
Golden jackal (*Canis aureus*)
Gorilla (*Gorilla* spp.)
Grant's gazelle (*Nanger granti*)
Gray whale (*Eschrichtius robustus*)
Gray wolf (*Canis lupus*)
Great auk (*Pinguinus impennis*)
Great egret (*Ardea alba*)
Great tit (*Parus major*)
Grevy's zebra (*Equus grevyi*)
Grey crowned crane (*Balearica regulorum*)
Guadalupe Island caracaca (*Caracara lutosa*)
Guadalupe Island petrel (*Oceanodroma macrodactyla*)
Guam flying fox (*Pteropus tokudae*)
Guam rail (*Gallirallus owstoni*)
Haast's eagle (*Harpagornis moorei*)
Harpy eagle (*Harpia harpyja*)
Hartebeest (*Alcelaphus buselaphus*)
Hawaiian crow (*Corvus hawaiiensis*)
Hawaii creeper (*Oreomystis mana*)
Himalayan quail (*Ophrysia superciliosa*)
Hippopotamus (*Hippopotamus* spp.)
Hokkaido wolf (*Canis lupus hattai*)
Hyacinth macaw (*Anodorhynchus hyacinthinus*)
Hyena (Family Hyaenidae)
Iguaca parrot (*Amazona vittata*)
Iiwi (*Vestiaria coccinea*)
Imperial woodpecker (*Campephilus imperialis*)
Inca rat (*Cuscomys oblativa*)
Indian gazelle or chinkara (*Gazella bennettii*)
Indian rhino (*Rhinoceros unicornis*)
Island's unique caracara (*Caracara lutosa*)
Ivory-billed woodpecker (*Campephilus principalis*)
Jaguar (*Panthera onca*)
Jamaican fruit-eating bat (*Artibeus jamaicensis*)
Javan rhino (*Rhinoceros sondaicus*)
Javan tiger (*Panthera tigris sondaica*)
Jucara (*Euterpe edulis*)
Kabomani tapir (*Tapirus kabomani*)
Kakapo (*Strigops habroptilus*)
Killer whale (*Orcinus orca*)
King penguin (*Aptenodytes patagonicus*)
Kipunji (*Rungwecebus kipunji*)
Kit fox (*Vulpes macrotis*)
Kittiwake (*Rissa* spp.)
Kiwi (*Apteryx* spp.)
Large-leafed mangrove (*Bruguiera gymnorhiza*)
Large Palau flying fox (*Pteropus pilosus*)
Laysan albatross (*Phoebastria immutabilis*)
Lear's macaw (*Anodorhynchus leari*)
Leopard (*Panthera pardus*)
Leopard seal (*Hydrurga leptonyx*)
Lesser bilby (*Macrotis leucura*)
Lion (*Panthera leo*)
Little brown bat (*Myotis lucifugus*)
Little Marianas fruit bat (*Pteropus tokudae*)
Lowland gorilla (*Gorilla gorilla gorilla*)
Macaque (*Macaca* spp.)
Maclear's rat (*Rattus macleari*)
Manatee (*Trichechus manatus*)
Maned sloth (*Bradypus torquatus*)
Marianas flying fox (*Pteropus mariannus*)
Mauritian flying fox (*Pteropus niger*)
Merriam elk (*Cervus canadensis merriami*)
Mexican free-tailed bat (*Tadarida brasiliensis*)
Mexican long-nosed bat (*Leptonycteris nivalis*)
Mexican wolf (*Canis lupus baileyi*)
Micronesian starling (*Aplonis opaca*)
Moas (Family Dinornithiformes)
Moose (*Alces alces*)
Mountain gorilla (*Gorilla beringei beringei*)
Mountain plover (*Charadrius montanus*)
Mouse lemur (*Microcebus* spp.)
Muntjac deer (*Muntiacus* spp.)
Murphy's petrel (*Pterodroma ultima*)
Murres (*Uria aalge*)

Nepali hog plum (*Choerospondias axillaris*)
Night monkey (*Aotus* sp.)
Nilgai (*Boselaphus tragocamelus*)
North American beaver (*Castor canadensis*)
North Atlantic right whale (*Eubalaena glacialis*)
Northern buffed-cheeked gibbon (*Nomascus annamensis*)
Northern muriqui (*Brachyteles hypoxanthus*)
Northern right whale (*Eubalaena glacialis*)
North Pacific right whale (*Eubalaena japonica*)
Norway rat (*Rattus norvegicus*)
Orangutan (*Pongo* spp.)
Ornate-hawk eagle (*Spizaetus ornatus*)
Ostrich (*Struthio camelus*)
Oystercatchers (*Haematopus* spp.)
Pacific rat (*Rattus exulans*)
Pacific sheath-tailed bat (*Emballonura semicaudata*)
Pampas deer (*Ozotoceros bezoarticus*)
Paradise parrot (*Psephotus pulcherrimus*)
Passenger pigeon (*Ectopistes migratorius*)
Pelican (*Pelecanus* spp.)
Père David's deer (*Elaphurus davidianus*)
Peregrine falcon (*Falco peregrinus*)
Persian gazelle (*Gazella subgutturosa*)
Peruvian pygmy beaked whale (*Mesoplodon peruvianus*)
Philippine monkey (*Carlito syrichta*)
Philippine monkey-eating eagle (*Pithecophaga jefferyi*)
Pileated woodpecker (*Dryocopus pileatus*)
Pink-headed duck (*Rhodonessa caryophyllacea*)
Plains zebra (*Equus quagga*)
Pohnpei lorikeet (*Trichoglossus rubiginosus*)
Polar bear (*Ursus maritimus*)
Proboscis monkey (*Nasalis larvatus*)
Pronghorn antelope (*Antilocapra americana*)
Przewalski's horse (*Equus ferus przewalskii*)
Puerto Rican Amazon (*Amazona vittata*)
Puffin (*Fratercula* spp.)
Purple martin (*Progne subis*)
Quagga (*Equus quagga quagga*)
Range-headed parrot (*Pyrilia auriantocephala*)
Razorbill (*Alca torda*)
Red-billed oxpecker (*Buphagus erythrorhynchus*)
Red-browed Amazon (*Amazona rhodocorytha*)
Red-crowned crane (*Grus japonensis*)
Red fox (*Vulpes vulpes*)
Red giant flying squirrel (*Petaurista petaurista*)
Redshank (*Tringa tetanus*)
Reed-warbler (*Acrocephalus* spp.)
Resplendent quetzal (*Pharomachrus mocinno*)
Rhea (*Rhea* spp.)
Ringed plover (*Charadrius hiaticula*)
Rock dove (*Columba livia*)
Rowi (*Apteryx rowi*)
Sable antelope (*Hippotragus niger*)
Saiga (*Saiga tatarica*)
Sambar deer (*Rusa unicolor*)
Santa Marta mountain-tanager (*Anisognathus melanogenys*)
Santa Marta wren (*Troglodytes monticola*)
Saudi gazelle (*Gazella saudiya*)
Savanna elephant (*Loxodonta africana*)
Scarlet macaw (*Ara macao*)
Schomburgk's deer (*Rucervus schomburgki*)
Seven-colored tanager (*Tangara fastuosa*)
Short-tailed shearwater (*Puffinus tenuirostris*)
Siberian tiger (*Panthera tigris altaica*)
Side-striped jackal (*Canis adustus*)
Silver marmoset (*Mico argentatus*)
Skua (*Stercorarius* spp.)
Slate-colored grosbeak (*Saltator grossus*)
Slender-billed grackle (*Quiscalus palustris*)
Sloth bear (*Melursus ursinus*)
Slow loris (*Nycticebus spp.*)
Snowy egret (*Egretta thula*)
South American coati (*Nasua nasua*)
South China tiger (*Panthera tigris amoyensis*)
Southern cassowary (*Casuarius casuarius*)
Southern long-nosed bat (*Leptonycteris curasoae*)
Southern muriqui (*Brachyteles arachnoides*)
Southern plains gray langur (*Semnopithecus dussumieri*)
Sperm whale (*Physeter macrocephalus*)
Spix's macaw (*Cyanopsitta spixii*)
Starling (Family Sturnidae)
Steller's sea cow (*Hydrodamalis gigas*)
Stephens Island wren (*Xenicus lyalli*)
Stoat or short-tailed weasel (*Mustela erminea*)
Sumatran orangutan (*Pongo abelii*)
Sumatran rhino (*Dicerorhinus sumatrensis*)

Syrian wild ass (*Equus hemionus hemippus*)
Tambalacoque (*Sideroxylon grandiflorum*)
Tarpan (*Equus ferus ferus*)
Tasmanian devil (*Sarcophilus harrisii*)
Tasmanian tiger or thylacine (*Thylacinus cynocephalus*)
Thomson's gazelle (*Eudorcas thomsonii*)
Tiger (*Panthera tigris*)
Tokoeka (*Apteryx australis*)
Toolache wallaby (*Macropus greyi*)
Topi antelope (*Damaliscus korrigum*)
Toucan (Family Ramphastidae)
Truk monarch (*Metabolus rugensis*)
Truk white-eye (*Rukia ruki*)
Tuamotu sandpiper (*Prosobonia cancellatus*)
Vaquita (*Phocoena sinus*)
Wattled crane (*Bugeranus carunculatus*)
West African black rhino (*Diceros bicornis longipes*)
Western lowland gorilla (*Gorilla gorilla gorilla*)
White-eyed river martin (*Pseudochelidon sirintarae*)
Whip-poor-will (*Antrostomus vociferus*)
Whistling thorn tree (*Vachellia drepanolobium*)
White-backed vulture (*Gyps africanus*)
White-faced capuchin monkey (*Cebus capucinus*)
White-footed mouse (*Peromyscus leucopus*)
White-nosed coati (*Nasua narica*)
White rhino (*Ceratotherium simum*)
Whooping crane (*Grus americana*)
Wilson's storm petrel (*Oceanites oceanicus*)

Recommended Reading

We have selected and annotated a few important books, papers, and videos, which will supplement material included in this volume or simply give you some pleasurable reading. If you want further information on a topic, or to check our sources, then web search engines, such as Google Scholar, are now so good at that task that we do not need to burden you with a long list of technical references.

Preface

Ceballos, G., A. García, P. R. Ehrlich. 2010. The sixth extinction crisis: Loss of animal populations and species. *Journal of Cosmology* 8: 1821–1831. An overview of new discoveries and the current extinction crises in vertebrates.

Dirzo, R., H. Young, M. Galetti, G. Ceballos, N. J. Issac, B. Collen. 2014. Anthropocene defaunation. *Science* 345: 401–406. An excellent overview of the causes and effects of current species and population losses in vertebrates.

Ehrlich, P. R., A. H. Ehrlich. 2009. *The Dominant Animal: Human Evolution and the Environment*. 2nd ed. Island Press, Washington, D.C. Broad overview of the impact of human beings on the environment.

Flannery, T. 2001. *A Gap in Nature: Discovering the World of Extinct Animals*. Atlantic Monthly Press, New York. A fine introduction to the many beautiful and unusual animals that are no longer with us. Great illustrations.

Kolbert, E. 2014. *The Sixth Extinction: An Unnatural History*. Henry Holt, New York. A wonderfully written, almost poetic, description of the fate of biodiversity.

Chapter 1. The Legacy

Bodsworth, F. 2011 (1954). *The Last of the Curlews*. Counterpoint Press, Berkeley, Calif. Sad story, with afterword to the new edition by Murray Gell-Mann, Nobel Prize–winning physicist and demon birder.

Carroll, S. B. 2009. *Remarkable Creatures: Epic Adventures in the Search for the Origins of Species*. Mariner Books, Boston. Interesting essays on important scientists and their discoveries, including a nice overview of the Cretaceous massive extinction.

Daily, G. C. 1997. *Nature's Services: Societal Dependence on Natural Ecosystems*. Island Press, Washington, D.C. The fundamental source on ecosystem services.

Roberts, C. 2013. *The Ocean of Life: The Fate of Man and the Sea*. Penguin Group, New York. The best popular book on the state of the oceans. Contains information on topics ranging from the origin of life to the impacts of plastics and noise on the animals of the ocean.

Shucker, K. P. N. 2012. *The Encyclopaedia of New and Rediscovered Animals: From the Lost Ark to the New Zoo—and Beyond*. Coachwhip Publications, Greenville, Ohio. A fine compilation of the incredible array of new species of animals discovered or rediscovered in the past two decades.

Wilson, E. O. 1993. *The Diversity of Life*. Harvard University Press, Cambridge, Mass. Wonderful book by a great conservationist.

Chapter 2. Natural Extinctions

Leopold, A. 1966. *A Sand County Almanac, with Essays from Round River*. Ballantine Books, New York. Wonderfully written, gives a feel for the land not easily available.

MacLeod, N. 2013. *The Great Extinctions: What Causes Them and How They Shape Life*. Firefly Books, London. A fascinating history of natural extinctions.

Chapter 3. The Anthropocene

Diamond, J. M. 1997. *Guns, Germs, and Steel: The Fates of Human Societies*. W. W. Norton, New York. Prime source on how human cultures differentiated and how the modern world system (which is wiping out biodiversity) was formed.

Ehrlich, P. R. 2000. *Human Natures: Genes, Cultures, and the Human Prospect*. Island Press, Washington, D.C. The genetic and cultural evolution of our species.

Ehrlich, P. R., A. H. Ehrlich. 1981. *Extinction: The Causes and Consequences of the Disappearance of Species*. Random House, New York. Pioneering work.

Martin, P. S., R. Klein. 1989. *Quaternary Extinctions: A Prehistoric Revolution*. University of Arizona Press, Tucson. Classic work on the impact on biodiversity of human invasions.

Chapter 4. Long-Silenced Songs

Donald, P., et al. 2010. *Facing Extinction: The World's Rarest Birds and the Race to Save Them*. Poyser, London. An expert compilation of species accounts and conservation efforts.

Fuller, E. 2001. *Extinct Birds*. Comstock Publishing Associates, Ithaca, N.Y. A sad global overview.

Fuller, E. 2014. *Lost Animals: Extinction and the Photographic Record*. Princeton University Press, Princeton, N.J. Seeing photos of extinct creatures is an emotional experience for many of us.

Greenway, J. C. 1967. *Extinct and Vanishing Birds of the World*. Dover Publications, New York. Classic overview of characteristics, habitats, and other features of threatened and vanished birds.

Quammen, D. 1997. *The Song of the Dodo: Island Biogeography in an Age of Extinctions*. Scribner, New York. A wonderful read and an introduction to some of the most important ideas about how organisms are distributed, speciate, and go extinct.

Chapter 5. Birds in Trouble

Search here for "vanishing or threatened birds" to see current status of the world's avifauna, including all species: www.iucnredlist.org/.

Snyder, N. F. R, D. E. Brown, K. B. Clark. 2009. *The Travails of Two Woodpeckers: Ivory-Bills and Imperials*. University of New Mexico Press, Albuquerque. Well-done story of two of the most charismatic North American birds, both still reported alive but both very likely extinct.

Wilcove, D. S. 1999. *The Condor's Shadow: The Loss and Recovery of Wildlife in America*. W. H. Freeman, New York. Great, very readable, overview.

Chapter 6. Mammals Lost

List of mammals gone extinct in historic times: www.petermaas.nl/extinct/lists/mammals.htm#.

Allen, G. M. 2012 (1942). *Extinct and Vanishing Mammals of the Western Hemisphere, with the Marine Mammals of All of the Oceans*. Hardpress Publishing, New York. A wonderful historic account, now available in paperback.

Ceballos, G., ed. 2014. *Mammals of Mexico*. Johns Hopkins University Press, Baltimore. See especially introductory overview on extinctions in Mexico.

Hornaday, W. T. 1913. *Our Vanishing Wild Life*. New York Zoological Society, New York. A superb historical description of early efforts in the United States to save endangered mammals such as the bison.

Johnson, C. 2007. *Australia's Mammal Extinctions: A 50,000-Year History*. Cambridge University Press, Cambridge, U.K. A well-written overview of mammals' extinctions in Australia, the continent where most mammal historic extinctions have occurred.

Tuvey, S. 2009. *Witness to Extinction: How We Failed to Save the Yangtze River Dolphin*. Oxford University Press, Oxford. An account of the sad recent extinction of a unique freshwater dolphin.

Chapter 7. Vanishing Mammals

Search here for "mammals" to see current status of all the world's mammal species: www.iucnredlist.org/.

Estes, J. A., et al., eds. 2006. *Whales, Whaling, and Ocean Ecosystems*. University of California Press, Berkeley. Combine with Callum Roberts's book (above) if you are interested in the plight of marine mammals.

Matthiessen, P. *Tigers in the Snow*. 2001. North Point Press, New York. The story of the highly endangered Siberian tiger, told by a superb writer.

Chapter 8. Why It All Matters

Carson, R. 1962. *Silent Spring*. Houghton Mifflin, Boston. Still worth reading after all these years, and despite continuous attacks by the hirelings of the pesticide industry.

Wotton, D. M., D. Kelly. 2011. Frugivore loss limits recruitment of large-seeded trees. *Proceedings of the Royal Society B* 278: 3345–3354. An excellent experiment showing that concern for the persistence of plants that lose their avian seed-dispersers is justified.

Chapter 9. Drivers of Death

If you want an overview of the climate disruption, much excellent and dependable information can be found at SkepticalScience: www.skepticalscience.com/.

Barnosky, A. 2009. *Heatstroke: Nature in an Age of Global Warming.* Island Press, Washington, D.C. Wonderful exposition from the viewpoint of a leading paleontologist.

Cribb, J. 2014. *Poisoned Planet: How Constant Exposure to Manmade Chemicals Is Putting Your Life at Risk.* Allen and Unwin, Crows Nest, NSW, Australia. Brilliant, scary account of a most important threat to biodiversity: the toxification of Earth. See especially the part on nanotechnology.

Klare, M. T. 2012. *The Race for What's Left: The Global Scramble for the World's Last Resources.* Metropolitan Books, New York. Will introduce you to the global resource situation, and you don't need to be a rocket scientist to see what it means for that other main driver, overconsumption by the rich.

Michaux, S. 2013. *Peak mining,* www.youtube.com/watch?v=TFyTSiCXWEE. This lecture brings home the message embodied in Klare's book.

Oreskes, N., E. M. Conway. 2010. *Merchants of Doubt: How a Handful of Scientists Obscured the Truth on Issues from Tobacco Smoke to Global Warming.* Bloomsbury Press, New York. Superb volume detailing the activities of well-organized and highly paid agents for unethical industries.

Weisman, A. 2013. *Countdown: Our Last, Best Hope for a Future on Earth?* Little, Brown, and Company, New York. Excellent recent overview of the population problem.

Chapter 10. Beyond Mourning

Diamond, J. 2005. *Collapse: How Societies Choose to Fail or Succeed.* Viking, New York. Best popular book on what can happen to societies when they get into ecological trouble.

Ehrlich, P. R., A. H. Ehrlich. 2013. Can a collapse of civilization be avoided? *Proceedings of the Royal Society B,* http://rspb.royalsocietypublishing.org/content/280/1754/20122845. What needs to be done to prevent a collapse that might well destroy much of the remaining bird and mammal faunas.

Ehrlich, P. R., R. E. Ornstein. 2010. *Humanity on a Tightrope: Thoughts on Empathy, Family, and Big Changes for a Viable Future.* Rowman & Littlefield, New York. Further thoughts on how to change the direction of society.

Kareiva, P., et al., eds. 2011. *Natural Capital: Theory and Practice of Mapping Ecosystem Services.* Oxford University Press, Oxford, U.K. A nice overview of ecosystem services.

Tainter, J. A. 1988. *The Collapse of Complex Societies.* Cambridge University Press, Cambridge, U.K. Best technical book on how to detect when a collapse is coming.

Index

Photography and Illustration Credits

Images by the following photographers appear on the pages listed: Scott Altenbach, from *Los Mamiferos Silvestres de Mexico,* Fondo de Cultura Economica—Conabio, Mexico D.F., 120, 130; Bristol City Museum / NPL / Minden Pictures, 75; Gerardo Ceballos, 10, 15, 18–19, 22, 23, 25, 26–27, 54, 78–79, 88, 101, 105, 109, 113, 122, 125, 134, 143, 146, 152, 165; Gerardo Ceballos and Paul R. Ehrlich, 34, 37, 39; Claudio Contreras Koob, 12–13, 28–29, 52, 60, 82; Paul R. Ehrlich, ix, 140–41, 152, 167, 168, *left;* Chris Fertnig / Thinkstock, 66–67; Peter Harrison, Apex Expeditions, 168, *right;* John Hessel, 127, 171; Jack Jeffrey Photography, 45, 121, *top and bottom,* 153; Frans Lanting / www.lanting.com, 2–3, 21, 46–47, 56, *top and bottom,* 58, 63, 74, 90, 99, 103, 106, 108, 116–17, 137, 144, 149, 151, 160–61, 176; Carl Lumholtz, from *Las Aves de México en Peligro de Extinción,* Fondo de Cultura Economica, Mexico D.F., 36; Susan McConnell, 7, 8–9, 16, 77, 87, 94–95, 97, 163; New York Zoological Society, 72; Alexander Pari, 5, *top;* Roberto Quispe, 5, *bottom;* Roland Seitre / NPL / Minden Pictures, 68; Lynn M. Stone / NPL / Minden Pictures, 51; Jorge Urbán, 80.

Illustrations by Ding Li Yong appear on the following pages: 32, 40, 42, 70, 71.

Jacket images, *left to right, top to bottom:* scarlet macaws (Frans Lanting); tiger (Gerardo Ceballos); chimpanzee (Susan McConnell); Indian rhino (Gerardo Ceballos); gray whale (Claudio Contreras Koob); whooping crane (Lynn M. Stone / NPL / Minden Pictures); polar bear (Frans Lanting); indri (Susan McConnell); giant panda (Frans Lanting); black rhino (Susan McConnell); chital (Gerardo Ceballos); ornate hawk-eagle (Claudio Contreras Koob); kiwi (Frans Lanting); lowland gorilla (Susan McConnell); aya-aye (Frans Lanting); baiji (Roland Seitre / NPL / Minden Pictures).